Disaster on Mount Slesse

DISASTER
on
MOUNT SLESSE

The Story of Western Canada's Worst Air Crash

Betty O'Keefe & Ian Macdonald

CAITLIN PRESS

Caitlin Press Inc.
P.O. Box 219
Madeira Park, BC V0N 2H0
www.caitlin-press.com

Cover design and maps by Warren Clark
Printed and bound in Canada

Caitlin Press acknowledges financial support from the Government of
Canada through the Book Publishing Industry Development Program and
the Canada Council for the Arts, and from the Province of British Columbia
through the BC Arts Council and the Book Publishing Tax Credit.

THE CANADA COUNCIL | LE CONSEIL DES ARTS
FOR THE ARTS | DU CANADA
SINCE 1957 | DEPUIS 1957

BRITISH
COLUMBIA
ARTS COUNCIL
Supported by the Province of British Columbia

Library and Archives Canada Cataloguing in Publication

O'Keefe, Betty, 1930–

Disaster on Mount Slesse : the story of Western Canada's worst air
crash / Betty O'Keefe and Ian MacDonald.

Includes bibliographical references and index.
ISBN 1-894759-21-4

1. Aircraft accidents—British Columbia—Slesse Mountain Region—
History. I. Macdonald, Ian, 1928– II. Title.

TL553.53.C3O34 2006   363.12'4650971137   C2006-903358-7

This book was written in honour of the passengers and crew of Trans-Canada Airlines Flight 810 who died on the evening of December 9, 1956, in what was then the country's worst air disaster.

Contents

Acknowledgements

The information about this Canadian tragedy came from official documents and files, newspaper accounts and a host of individuals who took the time to explain their personal involvement.

The authors are particularly indebted to Jay Clarke, son of Flight 810's pilot, and to other members of the Clarke family who provided invaluable information about Jack Clarke and his wife Vivian.

Our thanks also goes to Bob Brown with the Civil Air Search and Rescue Association in Chilliwack, whose enthusiastic, untiring assistance in obtaining facts from many local people was crucial to the book, as were the details provided by Max Abrams and the veterans of the Royal Winnipeg Rifles, the "Little Black Devils," who still meet regularly to remember one of their own who died on the flight.

We would also like to thank local residents who responded to a request for information that appeared in the *Chilliwack Progress* in 2005. Your comments and emails were sincere and personal. Included in this group were George Hill, Carolyn Paisley of the Chilliwack Public Library, Laura and Sig Peters, Corky Wiens, Fred and Rachel Bryant, Dorothy Baker, Verny Nelson and numerous others.

Valuable assistance also was provided by our friend and colleague Paddy Sherman (and his book *Cloud Walkers*, Macmillan 1965); Ruth Casey of Transport Canada, Pacific Region; Dick Jennens at Nav Canada; retired Air Canada pilot Jim Griffith; and members of the Sturtridge and Borthistle families, descendants of Saskatchewan Roughriders player Gordon Sturtridge, who died with his wife on the fatal flight.

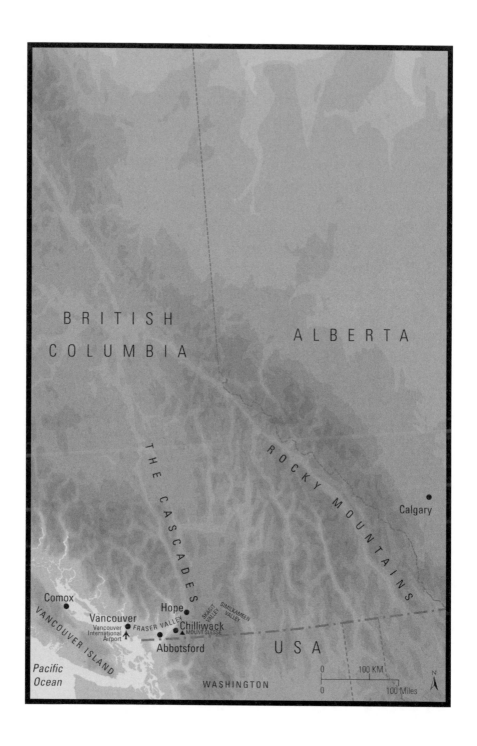

B R I T I S H
C O L U M B I A

A L B E R T A

T H E   C A S C A D E S

R O C K Y   M O U N T A I N S

Calgary

Comox

Hope

SKAGIT   SIMILKAMEEN
VALLEY   VALLEY

Vancouver
Vancouver   FRASER VALLEY
International   Chilliwack
Airport   MOUNT SLESSE

USA

Abbotsford

Pacific
Ocean

WASHINGTON

0   100 KM

0   100 Miles

N

11

Introduction

E ver since daring aviators first took to the skies over British Columbia early in the twentieth century, the province's great expanse of green mountains, fog-enveloped coastline, frozen peaks, deep river valleys and lakes has swallowed up numerous aircraft. Even as recently as 2004, it took several months of searching before a downed cross-country glider was found in the wilds of the BC Interior—unfortunately with the pilot dead at the controls. The same terrain has claimed sturdy commercial and military aircraft flown by veteran pilots; many of these tragedies remain shrouded in mystery.

In December 1942, a twin-engine Canadian Pacific Airlines (CPA) Lockheed 14 with thirteen people aboard disappeared on a flight between Prince George and Vancouver. Captain Ernest Kubicek, a veteran pilot who had flown Yukon mail routes in the 1930s, radioed Vancouver air traffic control to say that he was about fifteen minutes from the airport and that visibility was "just fair." That was the last message received. An extensive air search found no trace of the missing plane. On August 13 of the next year, however, as the snow melted off mountain peaks, an astute flyer spotted crash debris on Mount William Knight, part of the Cheam Range near Hope. A team of mountaineers led by Charles Woodworth, city editor of the *Vancouver Province,* reached the mountain about a week later. They scaled the mountain and confirmed that the debris was the remains of the CPA plane. It was neither the first nor the last plane to go down among the treacherous mountain ranges between Hope and the coast.

Another Vancouver-bound plane disappeared on April 28, 1947, again without a trace. The twin-engine Trans-Canada Airlines (TCA) Lodestar was carrying twelve passengers, two pilots and a stewardess on a direct flight from Lethbridge. As he neared the coast, Captain W.G. Pike radioed that he was flying at about 7,000 feet and would touch down in thirteen minutes. Winds were mild and though a light rain obscured views of the Vancouver runway, there were no reported landing difficulties. It was assumed that Captain Pike would take the plane about fifteen kilometres over the Gulf of Georgia and then turn 180 degrees east to swoop in over the water for a landing. Pike's message at 11:13 p.m. was the last anyone ever heard from the Lodestar. More than twenty planes—and all ocean-going vessels from tugboats to fishing boats—spent days scouring the Lower Mainland and nearby waters for signs of the Lodestar's wreckage. They found nothing. Nearly fifty years later, hikers stumbled upon the wreck on Mount Cheam near Chilliwack.

As planes became larger, so too did the tragedies. Trans-Canada Airlines introduced the sixty-two-passenger North Star into its fleet in 1946. It was a giant for its day and quickly became a company workhorse, flying tough cross-country and international routes with unparalleled reliability. After more than a decade of heavy use, TCA North Stars were involved in only two fatal mishaps. One of those was a tragic mid-air collision; the other remains the worst air disaster in the history of western Canada.

On December 9, 1956 during a regularly scheduled run from Vancouver to Calgary, Regina and points east, Trans-Canada Airlines Flight 810 lost an engine over Hope. The pilot decided to return to Vancouver. Minutes later, the radio became silent. The last message from the flight was transmitted at 7:10 p.m. The fates of the sixty-two people aboard would not be known for months.

A Spectacular—and Deadly— Landscape

The Cascade Mountains of British Columbia and Washington state stand as sentinels at the head of major river valleys in the Pacific Northwest. Airplane passengers often view these snow-capped peaks as they land at Vancouver International Airport; motorists see them through car windows as they head east out of the city and into the upper Fraser Valley along Highway 1. Just past the city of Chilliwack, the mountains rise steeply to become mighty peaks standing like giant granite guardians at the entrance to BC's Southern Interior.

The scenery is spectacular, but it's a dangerous landscape. In *British Columbia: A Natural History*, Richard Cannings and Sydney Cannings describe the geology of the area as a "hopeless complex . . .

a bad dream." In the history of the world, the Cascades are relative newcomers, their cataclysmic development having occurred just 100 million years ago. Other changes and upheavals since then have created some of the world's most impressive topography in a province where 75 percent of the land mass lies at 3,000 feet and above.

Larger than many countries, British Columbia sweeps from the shores of the Pacific Ocean to the Rocky Mountains, embracing great forests teeming with wildlife along the way. Its 25,000-kilometre coastline is a scenic wonder with fjords, inlets, bays and coves, much of it uninhabited. The rugged mainland coast is somewhat protected from Pacific storms by Vancouver Island and the Queen Charlottes.

Inland, the Fraser Valley is one of the largest and most fertile river basins in the country. Endless housing subdivisions now encroach on the farmland and sap much of the area's agricultural identity. But in the 1950s, the Fraser Valley was a quiet pastoral community with a sparse farming population. Chilliwack, one of the

The deadly "fangs" of Mount Slesse, photographed from the air. Photograph courtesy Dennis Nevett.

valley's larger towns, was the centre of the dairy industry. The cows that grazed its grasslands provided milk and butter for the homes of the Lower Mainland. The valley then held clearly defined towns and villages, each with a character all its own. Only one road snaked along the Fraser River on its way to Hope. There the road forked, one arm leading northeast along the tumbling Fraser Canyon, the other pushing its way east through the Cascades into the Skagit and Similkameen valleys and on to the broad Okanagan plateau.

The Cascades were largely uninhabited in the 1950s: only a few small ski lodges and sawmills dotted the landscape. The sport of skiing had not drawn many to the area and those who did venture in climbed the ski slopes on ramshackle rope tows and T-bar lifts. In the summer, hardy mountaineers dared the region's sheer cliffs and slushy peaks. Few braved them in winter. It was, and remains, a majestic land: awe-inspiring, unbelievably beautiful and frequently unforgiving—even to those who confront it with great care.

The Storm of '56

Vicious storms that brew over the Pacific Ocean frequently slam into southwestern British Columbia, often leaving a trail of devastation, injury and sometimes death. In early December 1956 a particularly severe storm carrying raging winds, ice, snow, sleet and rain tore through the Fraser Valley. For days it lashed the craggy Cascade Mountains at the valley head.

For weeks previous, life on Canada's southern Pacific coast was muted by fog and rain. Canals throughout the valley overflowed. With visibility only as far as the front hood, cars frequently missed curves and overturned in the water-filled ditches. A grey soup shrouded the entire region from the coast to Hope.

The wet, dreary month of November 1956 ended on a Friday, and fog closed the Vancouver Airport to all traffic for most of the day. In these days before radar and automatic pilot, flights were halted during bad weather. The unrelenting blanket descended again overnight, putting a damper on the week of preparations leading up to the December 8, 1956 Canadian Football League (CFL) All-Star Game that would mark the end of the season.

The sun finally shone again on Monday, December 3. People heaved a sigh of relief as they were able to travel freely. The reprieve was short-lived, however, for the next day brought a slushy snowfall that paralyzed motorists. An estimated 3,000 automobiles were stuck in a six-kilometre-long traffic jam leading up to the Lions Gate Bridge, linking downtown Vancouver to the city's North Shore. Forecasters predicted up to forty centimetres of new snow overnight.

Crews at Empire Stadium, where the big game would take place, stretched a tarpaulin over the field to protect it from the weather. Preparations continued unabated as locals kept their eyes skyward, hoping the rain would dry up for players and fans flying in from across the country. Instead, even before the official start of winter, Mother Nature ushered in an arctic front with temperatures of minus-seven degrees Celsius and winds of eighty kilometres per hour. By Thursday, sixty different hunters were stranded or missing in waist-deep snow near Clinton. There were even rare thirty-centimetre-deep drifts in Victoria. The cold front persisted for another day but then began to weaken. Winds dropped to forty kilometres an hour. Eight centimetres of snow were predicted for Saturday, the day of the All-Star Game. That morning, fans watched the snow change to sleet and then to driving rain.

The weatherman had predicted miserable weather, but nothing like the storm that swept in from the ocean on December 8 and 9 and tore across Vancouver Island and the Lower Mainland. It set records for sleet, snow and rain. Wind gusts of more than 125 kilometres per hour felled trees and power poles and ripped the roofs from buildings. The storm dashed madly up the valley, hurtling through mountain passes and over peaks and triggering mudslides on hills along highways and railway tracks. Victoria recorded 11.4 centimetres of rain in twenty-four hours; Abbotsford got 8.9. North Vancouver had never seen water cascading down the hills like this before; twenty centimetres fell in thirty-six hours. Municipal workers from every department, including garbage men and office staff, joined Boy Scouts, army cadets, housewives and businessmen on sandbag crews in an attempt to stem the flow of water. The water volume in drainage sewers along the lower slopes of North Vancouver mountains was so great that manhole covers popped off and slid down the steep slopes along with debris washed from basements, gardens and forests.

Chris Norberg, a visiting Californian, collapsed and died while fighting floods in New Westminster. Houses were flooded, drains became clogged and intersections turned into lakes so that many downtown streets became impassable. The cold snap was over and heavy rain would soon wash away the last traces of snow, but widespread flooding became the new problem.

An Ugly Football Game

The CFL All-Star game was held December 8, 1956 in Vancouver at the brand-new Empire Stadium, which had been built on the Pacific National Exhibition grounds in the city's east end to house the 1954 British Empire Games. The 32,375-capacity venue had gained instant worldwide fame when England's Roger Bannister and Australia's John Landy both broke the four-minute-mile barrier.

This would be the second annual East West Shrine All-Star Game in Canada. The first—so named because proceeds went to the Shriners—was held in Toronto in 1955, also in bad weather, and ended in an uninspiring 6–6 tie. Because it followed the CFL's showcase Grey Cup game by a couple of weeks, the 1956 All-Star was difficult for the Shriners to promote, even in a football-mad city like Vancouver. Although it was held in support of Shriners charities, the game sold only 17,000 advance tickets at $7.50 each.

The weather wreaked havoc with the big game. Many would-be spectators were delayed or gave up on their way to the stadium because their vehicles stalled in huge puddles. Some decided to stay home and avoid the flooding altogether. Nearly 4,000 ticket-holders failed to show up; only 13,546 splashed their way to the event to sit in the bitter cold. It was unfortunate timing for the game, which quickly became known as "the soup bowl."

Vancouver sports fans had apparently adopted the opinion that attending a sporting event in early December, when rain was almost guaranteed to come down hard and heavy, was not the best idea—especially at the open-air Empire Stadium. The stadium was far from a jewel. Only 10,000 seats were partly covered, providing little protection from the cold and rain for the three-hour game. Doleful Shriners cancelled plans for a parade of bands that had been planned as part of the half-time entertainment.

The Western Conference All-Star Team, 1956. BC Sports Hall of Fame.

The rosters of both sides battling in the 1956 All-Star tilt reads today like a who's-who of Canadian football in the '50s and '60s. Representing the Edmonton Eskimos were the legendary Jackie Parker and Normie Kwong. From other teams across the country came By Bailey, Sam Etcheverry, Bud Grant, Hal Patterson and Dick Shatto. The Saskatchewan Roughriders sent defensive end Gordon Sturtridge and centre Mel Becket. Two of their teammates, guards Mario DeMarco and Ray Syrnyk, came from Regina just to watch the game and sat glumly through the wet and wind—becoming colder and wetter as the game progressed. Calvin Jones, another guard, represented the Winnipeg Blue Bombers.

Frank "Pop" Ivy, whose Edmonton team had won the Grey Cup two weeks earlier, coached the Western team, while Jim Trimble from the Hamilton Tiger-Cats directed the Eastern squad. For sixty minutes, the teams waded back and forth across the sodden field. The West were the better swimmers, winning by a ridiculous 35–0.

In the *Province*, sports columnist Jim Kearney called the All-Star Game "an all-star flop," and added that spectators who risked pneumonia deserved something better. Unquestionably, the game's sharpest critic was the Ottawa Rough Riders' Don Phiney who blasted his eastern teammates for not taking the game seriously, saying they had come to party rather than play.

Most of the defeated eastern all-stars flew home early Sunday. The winning westerners had booked tickets back to the Prairies on various flights, one of which was Trans-Canada Airlines Flight 810, a milk run to Toronto with stops in Calgary, Regina and Winnipeg. It was scheduled to depart Vancouver International Airport at 4 p.m. on Sunday.

4

The Flight 810 Crew

J ust as it had wreaked havoc with the football game, the storm continued to leave a path of destruction as it moved east, disrupting travel plans across the country. On the afternoon of December 9 veteran pilot Captain Alan Jack Clarke could see only glowering, rain-sodden skies as he looked out the door of his Vancouver home, but he told his wife it was a beautiful day for flying. The thirty-five-year-old loved his job and all its challenges, even when he was called in to cover for a sick co-worker, as he was that day.

Clarke, a captain with TCA, was a native of Montreal but now flew mainly between Vancouver and Winnipeg. It was a busy Sunday for him and his family, who had just bought their comfortable home a year earlier. Jack and his wife, Vivian, thirty-four, were celebrating their eleventh wedding anniversary with sons Jay, nine, and Teddy, seven.

Captain Jack Clarke was called to cover Flight 810 for a sick co-worker. Used with permission of Jay Clarke.

Vivian (Murdoch) Clarke had trained as a registered nurse and, after working in a hospital for several years, decided to become an airline stewardess. She met Clarke while flying with TCA and they were married shortly thereafter.

Their anniversary celebration had been delayed a week because Clarke had been fogged in at Winnipeg the previous Sunday. He gave Vivian a ceramic drop necklace made by a local artisan, and she told the boys that the two masks in the design represented joy and tragedy. She explained to them that in everyone's life were times of great happiness and others of great sadness. She told them she hoped they would live long and have great joy despite inevitable times of sorrow. She did not know how prophetic her words would become.

The house was a happy jumble of written and unwritten Christmas cards and wrapped and unwrapped presents. Jack received a call from work informing him that his flight would be delayed. He left for the airport in the mid-afternoon, telling the boys to leave some wrapping work for him to help with when he returned. "It'll be a piece of cake," he said to Vivian as he waved goodbye. "I'll be back for dinner tomorrow."

Flight 810 would stop at Calgary, Regina and Winnipeg before carrying on to Toronto with a new crew. The plane, a Canadair North Star, had been late arriving from the east because of heavy headwinds and turbulence. Many of the outbound passengers had been waiting at the dreary terminal for hours and were getting impatient. While they waited, six of the passengers took out a total of $265,570 in flight insurance at a booth in the waiting room.

Clarke met up with co-pilot John Clarence (Terry) Boon and together they headed for a pre-flight weather briefing at 4:30 p.m. Trim, fit men, sharp in their well-pressed navy blue uniforms, they exuded the impression that company managers always hoped to convey to first-time flyers or fidgety passengers. They were the epitome of calm, confident professionals, men whom passengers could trust.

Inclement weather was nothing new for Clarke and Boon, both former Air Force flyers. Clarke had lived through the hell and horror of bombing raids over Europe from which many of his fellow flyers had not returned, and both had seen worse conditions more recently during winters at home.

Clarke was born in Montreal. He received a scholarship to the Southampton School of Art in Britain and trained there before returning to his hometown. He was working for Associated Screen News as an artist when the war began. In September 1940, as British Spitfires and Hurricanes were fending off the Luftwaffe in the Battle of Britain, the nineteen-year-old Clarke volunteered for service in the Royal Canadian Air Force (RCAF). After training at Fort William and Brantford, Ontario, he was shipped to Kinloss, Scotland in July 1941. Attached to No. 10 Squadron of the Royal Air Force (RAF), he

Jack Clarke was an accomplished Air Force flyer before becoming a commercial pilot. Used with permission of Jay Clarke.

flew out of bases in Yorkshire, England, first in Whitley aircraft and then in Halifax four-engine heavy bombers.

Clarke flew in the Battle of the Atlantic, including the "Channel Dash" on Hitler's battleships and in numerous raids against U-boat pens in France. He took part in the destruction of the warship *Gneisenau* at Kiel on February 26, 1942. In May and June he flew on raids to Cologne, Essen and Bremen, where he dodged searchlights, flak from anti-aircraft guns and night fighters defending Germany. He often returned with a bullet-ridden fuselage. After one raid, he sputtered back on only three engines. By July, No. 10 Squadron was posted to Egypt for six months, where Clarke took part in the Battle of El Alamein, the turning point of the African front. When his first tour was completed, he opted to stay with his crew in the RAF for a second tour rather than transfer to a Canadian squadron that was then being formed.

Clarke survived an astonishing forty-seven combat missions in all. He completed more operations than most of those in the Bomber Command, which led to his being released from the war zone early. He sailed to New York in 1944 and was posted to Boundary Bay south of Vancouver as an RCAF flight instructor on B-24 Liberator and B-25 Mitchell bombers.

In January 1945, Jack Clarke was loaned to TCA eight months prior to his formal discharge. He took TCA training in Winnipeg and was promoted to captain in June 1946. He was based in Lethbridge until October 1947, when he transferred to Winnipeg.

While there, Clarke became the commanding officer of No. 177 (TCA) Air Cadet Squadron. He was a dedicated leader who passed on his deep love of flight. Several students were inspired to become commercial pilots themselves, including Jim Griffith, who later flew for Air Canada for thirty-nine years. The young Griffith even pooled some money with his fellow cadets to buy Clarke a briefcase.

In May 1955, just before he was posted to Vancouver, Clarke

wrote a thank-you letter to Griffith that captures some of his enthusiasm for nurturing the young flyers:

> May I, through you, express to all the Officers, Non-Commissioned Officers, and Cadets of 177 (TCA) Squadron my heartiest thanks for the briefcase which you so kindly presented to me on the fitting occasion of the annual Father and Son Banquet on May 5.
>
> The briefcase will serve as a constant reminder of the many fine friendships which I have made among the squadron these past three and a half years and which will always occupy my thoughts—the "ups and downs" we have all had. Indeed it will know many more ups and downs of a different nature when it flies with me from Vancouver.
>
> I know that all of you had a part in this useful and befitting gift and this will serve to make me feel that on every flight "all of you" will be present in the cockpit.
>
> I would be remiss if I failed to mention the deep appreciation I have for all the excellent co-operation all ranks have shown me during my years of command and ask that this fine spirit be maintained as you will know a CO's [commanding officer's] job is not always the easiest one!
>
> Once again, many, many thanks gentlemen and with my best wishes for your future success both individually and collectively, I remain,
>
> Yours respectfully,
> A.J. Clarke.

Fifty years after Clarke's death, Griffith said of his mentor, "He was my role model and I felt that I owed him for giving me the opportunity to pursue an interesting, not to mention rewarding, career as an airline pilot."

For First Officer Terry Boon, age twenty-six, Flight 810 had particular significance. It was to be his last aboard the North Star before he began training to fly the newer British turboprop Vickers Viscount plane that was taking over many of the airline's domestic routes. Boon had joined TCA in July 1953. He was promoted to flying officer status three months later and assigned to the western region in July 1954. He qualified to fly the North Star in May 1955. His log showed he had nearly 4,000 hours flying with TCA, over 1,000 of which were at the controls of North Stars. The young pilot was a Vancouver native, unmarried and living on the city's North Shore. He had attended the University of British Columbia before joining the Air Force.

The third member of the crew was twenty-four-year-old stewardess Dorothy Elizabeth Bjornson, who came from the small agricultural community of Swan River, Manitoba. She had joined TCA in July after completing nursing studies at Winnipeg General Hospital, a nursing degree being a requirement for cabin crew at TCA and most other major airlines at the time. She was the only daughter in a family with six older brothers. Bjornson was responsible for the comfort and needs of the fifty-nine travellers about to board Flight 810. Keen on her new job, she, like Clarke, had agreed to fill in on the run for a sick co-worker.

At the weather briefing, Department of Transport airways forecaster D.N. McMullen filled Clarke and Boon in on exactly how bad the flying conditions were. When Clarke saw the weather graph, he quipped that the use of a black pencil wasn't bad enough; forecasters were also using red to show some of the extremely severe conditions. Weather for all of western Canada was awful. McMullen said the cloud cover started at 3,000 feet and extended to 20,000 feet. An earlier aircraft had radioed that there was moderate turbulence at 16,000 feet and severe turbulence below 10,000 in the area around Hope. Winds over the mountains reached 130 kilometres per hour at 15,000 feet and 160 kilometres per hour at 19,000. Clarke was told that because of local conditions he might find a westbound climb out of Vancouver more expedient than the usual eastbound one taken for a trip to Calgary. It was likely to be a bumpy ride for Flight 810's passengers.

The pilots also learned that the weather was no better farther east. There was thick cloud over the foothills of the Rockies near the BC–Alberta border, and the flight was expected to catch up with the same system that had dumped heavy rains on Vancouver earlier Sunday. Weather at Calgary and Edmonton could be so severe that the group might have to miss the first scheduled stop in favour of better weather at Regina, Saskatoon or Winnipeg. Warm, dry chinook winds were expected but wouldn't hit Calgary until after Flight 810's arrival. Captain Clarke was shown a weather pattern sketch for southern Alberta prepared by a TCA dispatcher showing the extent of the stormy skies. Only time would tell exactly what conditions would prevail and where the first stop would be made. An overflight would throw another wrench into the travel plans of the many Albertans aboard who planned to deplane at Calgary, and they were already well behind schedule because of the storm. It was not going to be an easy trip, but the pilots had few worries as they headed out on the tarmac toward the North Star.

5

The North Star

The Canadair North Star was the $680,000 workhorse of TCA's early fleet of about eighty aircraft. Design work had begun in 1944, most of it done at TCA engineering facilities in Montreal and Winnipeg in the days when Canada had a significant aircraft construction industry. Canadair had originally planned to construct forty-four planes—twenty-four for the Air Force and twenty-two for TCA. The fuselage was built by the Douglas Aircraft Company in Chicago and completed in Canada by Canadair to TCA and RCAF specifications. The selected design was essentially a Douglas DC-4 airframe with some DC-6 components and Rolls-Royce Merlin engines.

Canadair president Ben Franklin suggested the name North Star as a tribute to Polaris, guide to the traveller. A prototype took to the skies on July 15, 1946. From the beginning, passengers complained of excessive engine noise. Even with the introduction of quieter exhaust systems, travellers had to shout to converse. The first North Star was delivered to TCA on November 16, 1946. Dubbed "a giant" of the times, it had a cruising speed of 383 kilometres per hour and a top speed of 616 kilometres per hour. The plane went into transatlantic service on April 15, 1947, and became the favoured long-haul aircraft. In its first fifty-eight days of operation the North Star crossed the Atlantic 116 times. In all Canadair produced seventy North Stars, including four for Canadian Pacific Airlines and about twenty for British Overseas Airways Corporation. TCA president Gordon McGregor was proud of his much-loved North Star airliners' performance. When its flying days were over, McGregor often pointed out that the company had operated between twenty and twenty-three of the aircraft for fifteen years, and had flown more than 3.5 billion passenger miles.

The Canadair DC-4M2 North Star was a giant of the times. Canadian Museum of Aviation.

The first TCA fatality in a North Star occurred on April 8, 1954, near an Air Force training base in Moose Jaw. At the time there had been complaints from civilian pilots that some of the military's fledgling flyers were too daring with their newly acquired skills, and on more than one occasion had come dangerously close to civilian aircraft. The Canadian Air Line Pilots' Association had sent a letter to the federal Department of Transport (now Transport Canada) complaining about the number of buzzing incidents. Then, on a clear blue spring day with unlimited visibility, a yellow Harvard trainer with two aboard clipped the wing of a North Star passenger carrier. The airliner was ripped open and spun to the ground in flames. Horrified spectators in Moose Jaw watched as wreckage and bodies tumbled to the earth. All thirty-two passengers and four crew members died instantly. Several houses were hit by flaming debris; one woman died when her home was set ablaze. The Harvard seemed to be under control in the moments immediately after the crash, but it burst into flames when it tried to land on a nearby golf course. The two RCAF flyers aboard were also killed.

By 1956, there were some eighteen North Stars in TCA service on Canadian routes. The veteran aircraft had proved safe and reliable, but their days were numbered. They were being replaced by the growing popularity and importance of the jet engine, which quickly took over civilian service. TCA already had an order in for four Douglas DC-8s, the first of many four-engine jetliners that the company would acquire.

The old North Star departing from Vancouver on December 9, 1956, was configured with sixty-two economy seats, in two cabins, four abreast with a single centre aisle. With sixty-two passengers and crew, 1,700 gallons of gas and 971 pounds of mail and freight—including 27 bags of letters, 90 pounds of express parcels, flowers, television film and electrical equipment—Flight 810 would be fully loaded when it finally took off.

Vancouver International Airport in 1956 was a much smaller, more modest place than the world-class facility it is today, encompassing little more than a small control tower, a few offices, a coffee shop and a modest passenger terminal. The airport was in its infancy then and was located in what today is known as the south terminal, a portion of the airport now dedicated to small regional airlines. A sprawling collection of giant buildings now houses terminals, enormous baggage conveyor systems, hangars, hotels and parking facilities. Food concessions, restaurants, shops and bars cater to millions of passengers a year. Today's more northerly site was not opened until 1968. It has been expanded a number of times since then and continues to grow at an astounding rate as one of the major gateways between North America and Asia.

Anxious from the Outset

Fate, as always, had thrown together a passenger list that was a cross-section of life. Those boarding Flight 810 included football fans and players, soldiers, World War II veterans, a Mountie, three women returning home from a Hawaii holiday and several young businessmen, many of whom were the fathers of small children.

The travellers on December 9 were anxious to be on their way. There wasn't much to do in the small airport lounge. Even the available coffee wasn't great. A few read to pass the time. Several others paced back and forth peering into the murky skies, the nervous among them concerned about the terrible storms that had been battering BC. Some worried about missed connections and late arrivals. Some took the philosophical view that delays were a given when flying in Canada during the winter. The long wait encouraged conversation that covered everything from the previous day's football game to the unsettling topic of the weather.

Football fans among those waiting for Flight 810 would have instantly recognized the five CFLers in the lounge: Mario DeMarco, Mel Becket, Gordon Sturtridge, Ray Syrnyk and Calvin Jones.

It would have been hard to miss the gregarious Mario DeMarco, a boisterous twenty-eight-year-old guard who was prepared to talk to anybody, particularly if the person liked football. Fans among his fellow travellers were quick to get his views on the recent one-sided contest. DeMarco was a big, fun-loving graduate of Miami University who had been named a Canadian all-star in 1952 and 1954. He had made this trip to Vancouver only to cheer on his teammates from the stands. He had played briefly in the National Football League and with Montreal in the CFL before moving to Regina. Very popular with his teammates, DeMarco was a joker whose lively presence in the dressing room as well as on the field made him a key player for the Roughriders. A native of New Jersey, he was planning to apply for Canadian citizenship. DeMarco was married to an Edmonton nurse. They had a one-year-old son named Danny.

DeMarco ran a gas station in Regina with Mel Becket, his twenty-seven-year-old teammate who had slogged through the swampy game at Empire Stadium. Becket was from Chicago and had graduated from Indiana University. He had been playing as an end for Saskatchewan for four years and was particularly eager to get home that afternoon. His wife was due to give birth to their first child sometime before Christmas. Frank Bien, an employee at the gas station owned by the players, said he had spoken to Becket before he left for Vancouver and told him not to get injured because his baby was due and there were a lot of Christmas parties to attend. Bien remembered that Becket replied, "This is one time I'm going to take it easy. You will not see me jumping on them very hard this time."

Also joining the football discussion would have been Gordon

Mildred and Gordon Sturtridge in front of their home in Winnipeg, just after their marriage. They were both killed in the crash. Photograph courtesy Valerie Borthistle.

Sturtridge, a twenty-seven-year-old Winnipeg native who had played for Regina for five years. A popular and dependable defensive end, he had been named an all-star three times and had been a standout in the first Shrine Bowl game in Toronto. Sturtridge had played his usual dependable game in the mud and the downpour at Empire Stadium. He had brought his wife, Mildred, along to give her a welcome reprieve from their three small children at home. During the delay, Sturtridge phoned the babysitter looking after Vicki, six, Val, five, and Gordon Jr., fifteen months, explaining that he and his wife would be several hours late.

Young Ray Syrnyk was the Roughrider who accompanied DeMarco to Vancouver as a spectator. Twenty years old, he was a student at the University of Saskatchewan and a guard for the football team. He had gone to school in Saskatoon and was a member of the Saskatoon Hilltops, which won the 1953 Canadian Junior Football Championship. The young bachelor was happy to discuss football at any time and regaled the airport crowd with tales of games won and lost.

The last football player awaiting Flight 810 was Calvin Jones, twenty-three, an unmarried 225-pound guard with the Winnipeg Blue Bombers. An American, he had been named lineman of the year as a member of the University of Iowa team in the National Collegiate Athletic Association's prestigious Big Ten conference. Jones had played a single season with Winnipeg and had just signed another contract for 1957. He had been out partying after the game and had slept in and missed his early morning flight to Winnipeg, so he was quick to pick up one of the cancellations on Flight 810.

Young Pat Rowan from Calgary was excited to be in the company of the football players. He had attended the game with his parents, Wally and Susan, and his five-year-old sister Yvonne. The opportunity to talk to his heroes was the topper to a wonderful trip. His father was a long-time TCA company man who had lived in Calgary for eighteen years managing the airline's operations. Active in community and civic affairs, he had recently been elected as a city alderman. Calgary mayor D.H. MacKay called Wally Rowan "an aggressive young man who was already well respected by the administration."

At least two other football fans sat among the impatient travellers. Prominent businessman Harold McElroy, fifty-six, was a long-time Regina football supporter who never missed a home game if he could help it. He was proud of his team and sang the praises of the Roughriders who played in the All-Star Game. He had included attendance at Empire Stadium with a business trip, as had the man assigned to the seat next to him, Gordon Kennedy, who was personnel manager with BA Oil Company and a prominent Shriner in Toronto. Kennedy had been unable to get a direct flight home so he decided to take the milk run back to Toronto.

Also among the several businessmen was Harold Cleven, who

Flight 810 Crew and Passenger List

CREW

Captain Alan Jack Clarke, Vancouver; First Officer John Clarence Terry Boon, North Vancouver; Dorothy Elizabeth Bjornson, Vancouver

PASSENGERS

British Columbia

Richard Theodore Custer, South Burnaby; Fred W. Edwards, Vancouver; John Edward Henderson, Vancouver; Kenneth Roy Laird, Vancouver; John David Lyall, Vancouver; Ian Hamilton MacBeth, West Vancouver; Ronald W. Mitchell, Vancouver; Robert John Muir, Powell River; John Archibald Munro, Vancouver; Duncan MacKenzie Stewart, Vancouver; John Struthers, Vancouver; Leslie Edward Webb, Vancouver; Arthur Lawrence West, Vancouver; Frank John Wright, Vancouver; Harold Edwin Wright, Vancouver

Alberta

Mabel Florine Adams, Calgary; Helen Phelps Chapman, Calgary; Harold Cleven, Calgary; Karl Warren Collett, Calgary; Philip Edwin Gower, Calgary; Margaret Jean Grant, Calgary; Jean Christian Hamilton, Edmonton; Robert Winslow Hamilton, Edmonton; Audrey Cameron Harper, Calgary; John Bennett Hemming, Calgary; Aline Litovchenko, Calgary; Harold Clarence McElroy, Calgary; James Milne McKay, Calgary; Edwin Stanley Pettitt, Calgary; Sarah C. Rose, Calgary; Patrick Rowan, Calgary; Susan Gail Rowan, Calgary; Walter Peter Rowan, Calgary; Yvonne Rowan, Calgary

Saskatchewan

Melvin Howard Becket, Regina; Wilfred Emde, Indian Head; Mario Joseph DeMarco, Regina; Gordon Henry Sturtridge, Regina; Mildred A. Sturtridge, Regina; Raymond Nicholas Syrnyk, Regina

Manitoba

Denise Marie Beernaerts, St. Boniface; Donald Arthur Holden, Winnipeg

Ontario

Clarence Gordon Kennedy, Toronto; Frances Eleanor Welch, Toronto

Quebec

Maxwell Sheppard Bailey, Mount Royal

Hong Kong

Cheng Sao Chen Low; Yuen Gar; Yuen Wah Yoon; Lee Wah Ying Yuen

Japan

Hatsuko Hashimoto Dong, Osaka

United States

Marion Lewis Bright, Fort Worth; Eliza Duncan Burt, San Francisco; Calvin J. Jones, Steubenville; Anthony Folger, Dallas; Wong Fook, New York; Georgina Kafoury, Portland; Kwan Song, New York; Russell Smith Stratton, Los Angeles; Joan Elizabeth Williams, San Francisco had moved from Winnipeg to Calgary in 1954 to become general manager of a corrugated box company. An avid Rotarian and ardent football fan, he had pulled a few strings to make the trip to Vancouver.

One young passenger likely hadn't the faintest idea who the football players were. He had been in Canada, with its bewildering language and customs, only a few hours. He was sixteen-year-old Yuen Wah Yoon, travelling alone en route from Hong Kong. He was scheduled to get off in Regina to see his mother and three sisters before joining his father, a bakery owner in Cobalt, Ontario. The awed youngster was aware that the five big men so many people had approached in the lounge were special, but he probably didn't know why. In the whirl of strange conversations and unusual surroundings, there was only one thing that filled his mind to the exclusion of almost everything else. In a few hours he would reunite with the family he had been apart from for two long years. He could hardly wait. Getting out of China had not been easy. The family had paid a lot of money to have the boy smuggled into Hong Kong. He was finally nearing the end of his long journey.

Several of the travellers had military connections. One wore a medal for heroism on his tunic and carried himself with particular assurance. He was Major Philip Edwin Gower, who had been awarded the Military Cross when he stormed Normandy with the Royal Winnipeg Rifles on D-Day. He had shown great courage in leading his regiment against dug-in German machine gunners. On the third day of the invasion, he was captured and spent almost a year in a prisoner-of-war (POW) camp before being freed. A half-century later, he would still be well remembered and respected by the survivors among the men he had led. He chatted in the waiting room with other veterans, but he too had one thing utmost in mind. When the major left Calgary twelve months earlier for a year's duty in Japan and Korea, he had said goodbye to his wife

Anne, their two sons and their daughter. Now there were five people eagerly awaiting his return. Baby Diane, just ten months old, had been born while he was overseas. He was desperately keen to see her. His wife had sent pictures and told of the baby's antics in loving letters. Gower's duty overseas had seemed long and at times lonely, similar in some ways to his many months in a POW camp, but he was nearly home again.

Brigadier Harold Wright of Vancouver would have recognized Gower's award. A sailor in the First World War and an artillery battery commander overseas in World War II, Brigadier Wright maintained his military interests as a reserve member in western Canada. The next day, Jack Wasserman's popular column in the Vancouver Sun stated that had it not been for the late arrival of an advertisement, the Sun would have carried a notice of Wright's appointment as the new vice-president of Canada Bakeries—but it was a promotion Wright would not be able to take.

An impromptu military reunion of veterans in the waiting room included Ian H. MacBeth of West Vancouver, who had served with the Canadian Signal Corps in the Second World War and was awarded the British Empire Medal for service in Italy. Now he was an up-and-coming young businessman, assistant branch manager and sales representative for an industrial supply company. He was heading for Edmonton, leaving behind his wife and two small children—one a four-month-old.

Another waiting former serviceman was John Munro, a navigator with the RCAF in the Second World War. He was an appraiser who had just come home from a business trip and was looking forward to a quiet pre-Christmas break. He had reluctantly agreed to a request from a colleague for help with an out-of-town job.

Many of the travellers might have glanced in envy at the tans three young women in the lobby were bringing home from their first trip to Hawaii. Audrey Harper, Jean Grant and Helen Chapman looked forward to delivering the gifts packed in their suitcases and telling friends and relatives in Canada about the golden sands of Waikiki.

Wilfred Emde, a young RCMP constable, was flying home to Indian Head, Saskatchewan. He had been with the Mounties for eight years and had just been transferred from Union Bay on Vancouver Island to Regina, the first time he would be working so close to home. His mother, Mrs. Fred Emde, had driven from Indian Head to Regina in order to meet her son's plane. She was already waiting at the airport when she learned that the flight would be delayed because of bad weather.

For others it was a much different story. Denise Beernaerts sat mournfully in the waiting room. A few months earlier the excited nineteen-year-old had left Manitoba to go out on her own to a job in Vancouver as a telephone operator. Then came the shattering news of her father's death and now she was headed to St. Boniface, Quebec, for the funeral.

Sarah Rose of Calgary was also in mourning. At her family's insistence, she had taken a break to visit her sister in San Francisco following the death of her husband. Her sister was now travelling back with her. A day or two later, an insurance agent would go to Rose's home but there would be no one there. He was delivering a $5,000 cheque, part of her late husband's life insurance settlement.

For Eleanor Welch, it had been a difficult trip to the West Coast. She had flown to Vancouver out of concern for her sick mother and was now returning to her home in Scarborough. She was, however, anxious to see her teenaged daughter Judy, who had recently won the 1956 Miss Ontario contest. Judy later recalled that when she had been declared the winner, her mother had dashed up onstage to hug her.

Conversations lagged as time wore on. Some of the people had been at the airport for nearly four hours. Night was falling on Vancouver and everyone's patience was running thin.

There were a few late cancellations for Flight 810. One was Jackie Parker, star quarterback for the Edmonton Eskimos, who had played a key role in the west's All-Star win. Rather than going to Edmonton before heading home, he had decided to visit his family in Knoxville, Tennessee. Parker would go on to play in other Grey Cup games during a brilliant career in the CFL with the Edmonton Eskimos.

Roughrider fullback Bobby Marlow had been scheduled to return to Saskatchewan with his teammates but cancelled his booking Sunday afternoon after deciding to stay in Vancouver until Monday. Harry McBrine, secretary of the Canadian Rugby Union, and Thomas Drope, president of the Regina Junior Rams Football Club, had also changed their flight plans. These four men, and others, made snap decisions they would remember for the rest of their lives.

It is not known which one of the athletes managed to purloin the East West Shrine banner after the game, but it somehow appeared in the airport waiting room. It may have been DeMarco who carried it as the boarding announcement was finally made, and the proud footballers walked across the tarmac and climbed up the stairs to the plane.

7

The Flight

Just before six o'clock, the plane was at last filled with fuel, loaded with luggage and freight and ready to fly. Rain glistened on the silver skin of the North Star under the airport lights as the passengers crossed the tarmac and climbed aboard. Before modern gates, walkways and security checks, passengers made their way from the Vancouver Airport terminal walking across blacktop to the boarding stairs, which opened from the mid-section door. It had been a cold, windy, wet walk, despite helpful TCA staff with umbrellas. Bjornson greeted passengers as they entered the cabin, directing them to their seats and making sure each was settled and buckled in. Only three seats remained vacant. There were fifty-nine passengers aboard who heard the North Star's four motors pick up their noisy beat.

Flight 810 was two hours behind schedule when the plane finally

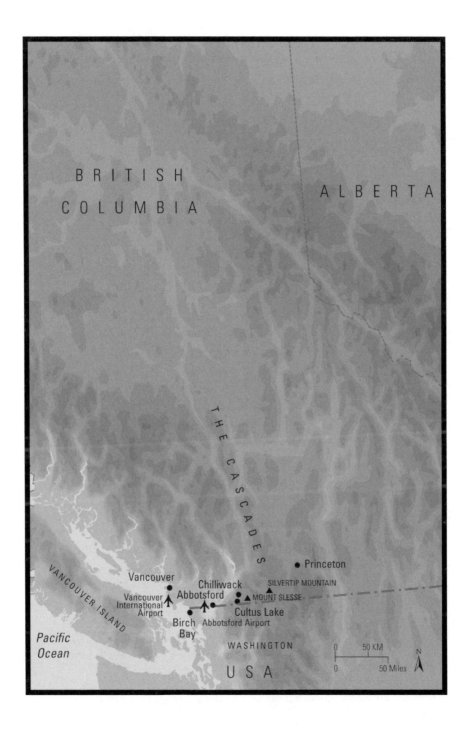

taxied toward the runway with Clarke and Boon at the controls. At 6:10 p.m. on December 9, 1956, Captain Clarke lifted the North Star off the runway. He headed west as he had been advised to do by air traffic control. The plane flew over Westham Island—just south of the airport at the mouth of the Fraser River—and then shuttled southward, climbing to 10,000 feet as it arced to the east.

The plane was assigned to fly the Green 1 air lane, a thirteen-kilometre-wide corridor that was the standard flight path to Calgary. Radio stations and signal-emitting beacons along the flight path transmitted navigational information that the crew would follow through BC and across the mountains. The signals from the radio ranges overlapped; when one started to fade the next came into range. By calculating their position in relation to each radio beacon, pilots could gauge where they were in all manner of dark and inclement weather. This was state-of-the-art flying technology in 1956.

Three minutes into the flight, Clarke had a radio conversation with his friend and colleague Captain Jack Wright, another TCA veteran, who was piloting Flight 7, a Super Constellation bound for Vancouver on the last leg of a flight from Toronto. Wright informed Clarke that he had experienced icing conditions over the Cascades and advised that it would be a good idea to fly at 19,000 to 20,000 feet going east. Clarke thanked him for the suggestion and asked Vancouver air traffic control for permission to climb to 21,000 feet.

At 6:21 p.m. Clarke received clearance for the flight to Calgary. He would fly through the Fraser Valley over Cultus Lake and Chilliwack before entering the mountains.

Some of the more inexperienced passengers in the cabin must have thought the flight's bumping and swaying seemed more extreme than the captain's casual description of "some turbulence." Stewardess Bjornson was strapped into her jump seat and unable to provide coffee or any other service to the passengers, some of whom clung to the arms of their chairs to steady themselves against the jostling of the aircraft. The loud roar of the North Star's four engines irritated even seasoned flyers, but perhaps it was reassurance to the nervous ones that the plane had adequate horsepower to lift them above the wind-packed storm and into smoother skies.

Fourteen minutes later the North Star reached 13,000 feet over Abbotsford; five minutes after that, the altimeter read 15,000 feet over Cultus Lake. At 16,000 feet Clarke radioed air traffic control to report light to moderate turbulence. He also indicated that icing was evident but not serious. The North Star's proven record over the years all over the world in some very bleak conditions had long before confirmed that the aircraft could handle these kinds of situations.

Clarke also let air traffic control know there had been sharp jolts between 16,000 and 18,000 feet, probably because he had flown through the tops of some cumulus clouds. At 6:48 he asked for clearance to climb to 21,000 feet, where Flight 810 might find smoother weather for the run over the Rockies.

Four minutes later Flight 810 reached 19,500 feet. The turbulence eased. Forty-two minutes into the flight, passengers finally began to relax.

Suddenly Clarke's instrument panel flashed a red warning light. He reacted immediately and shut off the No. 2 engine, the inner left-hand motor. He was back on the radio to air traffic control with an unexpected message: "Looks like we had a fire." The panel light could have been an indication of fire or a short circuit. There had been no visible signs of smoke or flames from the flight deck, but Clarke cut the engine anyway. He was well aware that electronic malfunctions in the instrument panel were not unknown in the North Star.

The captain did not hesitate to make his next move. He swung the plane to starboard 180 degrees, bearing back to the southwest.

Calmly, he asked air traffic control for altitude clearances for the return flight to Vancouver, stating he was at 19,000 feet and losing altitude fast. Flight 810 was cleared immediately to maintain 14,000 feet on Green 1.

As the plane completed the turn, Clarke radioed that he was now on Green 1 returning to Vancouver. But at a US military station in Birch Bay, Washington, a blip on a radar screen indicated that Flight 810 was well south of the normal air lane and had been off course from the time it swung around for the return flight. It disappeared from the screen near Silvertip Mountain. The radar station was not responsible for monitoring Canadian traffic, and the operator was unaware of the radio messages and the possible emergency unfolding. For the time being he ignored the blip. Vancouver air traffic control had no radar covering this area. It was unaware of the craft's actual position and unable to issue a course correction.

Several passengers on the left side of the plane could see that the engine closest to the fuselage was feathered, and everyone—except for a few seasoned flyers who had dozed off—must have felt the plane change direction. There was also a very noticeable increase in noise aboard the aircraft as the pilots throttled up the three working engines to maintain altitude.

Aware of the apprehension growing among his passengers, Clarke would have reassured them and explained that there was a problem with one of the engines that would necessitate a return to Vancouver. Undoubtedly, he apologized for the delay and inconvenience. It would have been difficult to hear the passengers' groans above the noise of the engines. Bjornson would have checked seat belts and answered any shouted questions as best she could. Still, a tense feeling must have pervaded the cabin.

At 7:01, air traffic control asked the pilot if he would be able to maintain 14,000 feet. "I think so," Clarke replied. According to the plane's altimeter, Flight 810 was now down to 15,000 feet. Controllers said the tone of Clarke's voice was calm. Another pilot, Captain C.L. Rickard, was flying Flight 4, a TCA Viscount east out of Vancouver. He was experiencing strong winds and moderate icing between 11,000 and 13,000 feet. He advised Clarke to maintain as much altitude as he could.

Just before reaching Hope, Clarke relayed a message from air traffic control to the Viscount, which was experiencing poor radio communications with Vancouver due to atmospheric conditions. Clarke had been worrying about the chances of a mid-air collision with the Viscount, but from the exchange he knew that the Viscount was well past his location and any chance of collision was over. He told Captain Rickard that he was being hit with heavy precipitation.

At 7:10 the North Star stated it was past Hope and requested clearance for a descent to 10,000 feet. Air traffic control cleared Flight 810 and told Clarke to cross the city at 8,000 or above. He acknowledged the message.

Any plane flying on Green 1 would soon clear the highest peaks of the towering Cascade mountain range. Vancouver Airport was only about 160 kilometres away. A secondary airport at Abbotsford was even closer and was suitable for an emergency landing. Under anything but the most unusual circumstances—and perhaps these conditions were more severe than most—the North Star was capable of three-engine flight without much difficulty. Clarke and his co-pilot were well trained to deal with what seemed to be a minor mishap. Flight 810 had plenty of fuel. It was estimated that only about 300 gallons of the 1,700 loaded at Vancouver had been expended. Aside from the lone radar blip in Washington state, there was nothing to suggest the plane wasn't flying along Green 1, clear of the mountain peaks at the head of the Fraser Valley. Only if Clarke had strayed off-course would the towering tops of the jagged range have been a danger.

Clarke's acknowledgement at 7:10 p.m. was the last message ever received from the North Star. In Vancouver, controllers became increasingly anxious as their repeated radio messages failed to raise the pilots. Somewhere out there in the blackness they knew that the partially crippled North Star was battling a violent storm while trying to reach the lights of the Vancouver runway.

The silence became increasingly ominous. Within a few minutes the controllers realized something could be gravely wrong. They set emergency measures in motion and let rescue crews know that an airliner with sixty-two people aboard was overdue, its whereabouts unknown.

The Search Begins

With each passing minute, air traffic controllers became more concerned about the plane's fate. They continued to scan the end of the runway, hoping to see Flight 810 come out of the night. There was a possibility that something had simply gone wrong with the radio and the plane would suddenly appear. It didn't.

It was only a few minutes after air traffic control failed to get an answer from the North Star that the controller alerted crash crews and firemen to the possibility that a crippled aircraft might be making an emergency landing. Trucks with red lights flashing immediately sped out along the runway.

At 8:00, air traffic control contacted Vancouver Search and Rescue headquarters with the information that radio communication had been lost with TCA Flight 810 somewhere in the vicinity of

Hope. The aircraft should have returned to Vancouver by 7:40, but there was still no sign of it. The RCAF's No. 121 Search and Rescue Squadron headquarters was at RCAF Station Jericho Beach along Vancouver's English Bay, within sight of Lions Gate Bridge and the downtown area of the city. Calls from the airport triggered action, and the squadron's log book detailed everything that followed.

The man on duty at the round-the-clock Search and Rescue recorded, "ATC [air traffic control] reports Trans-Canada Airlines' Flight 810 definitely overdue. Aircraft is a North Star." A notation in the book stated the plane had enough gas to stay airborne until 1:00 a.m. and was capable of flying even with two engines out, although veteran pilots knew that this would not keep a North Star airborne for long.

If Captain Clarke was forced to make an emergency landing, Abbotsford would have been the most likely alternative airport. Preliminary checks revealed there were no reports of an emergency landing there or at any other nearby field. Search and Rescue immediately broadcast a preliminary missing plane alert that would bring a response from all nearby airfields if any plane had made an unscheduled landing. There was still nothing, no reports of an emergency, no resumption of radio messages from Flight 810. News spread quickly to other search and rescue personnel. Off-duty staff began to make their way to Jericho. They wanted to help if they could.

So began the most intensive air search in Canadian history in what remains the worst plane disaster in western Canadian history. If all aboard were lost, it would be one of the ten worst plane crashes in the world at that time.

A pilot and navigator aboard a CF-100 all-weather fighter from the RCAF base at Comox on Vancouver Island were conducting an airborne training exercise when they were instructed to fly from Hope to Vancouver along the Green 1 air lane. The night sky was black, heavy with thunderheads and precipitation, but there was a possibility that fire or flares might be seen through a break in the thick cloud cover. Aircraft flying in the area had observed a few brief stretches of broken cloud. The CF-100's two-man crew spotted nothing as they flew as low as possible and peered down through the murky skies.

Other military aircraft were brought in from Vancouver Airport and RCAF Comox, which was then a major player in Canada's contribution to the North American air defence setup and home of an interceptor squadron equipped with the latest fighter aircraft. Called on were two DC-3s; another CF-100, a converted World War II Lancaster bomber still in use then by the RCAF for maritime patrol off the West Coast; and a Canso flying boat, another World War II veteran.

The second search plane to take to the air after the initial westbound flight by the CF-100 left Sea Island at 10:53 p.m., about three hours after air traffic control announced the plane overdue. It was a DC-3 with Flight Lieutenant Phil Walker at the controls. Fifteen minutes later another DC-3, flown by Flying Officer R.C. Tomlinson, took to the search. They were joined in the hunt by a Canso flying boat with Flying Officer George Waugh in command, and after midnight by a Lancaster from Comox. These crews could find no sign of fire or flares anywhere in the mountains between Hope and the US border as they scanned the ground through infrequent breaks in the cloud cover.

From the outset, the RCAF, TCA and other aviation and emergency organizations knew they could be engaged in the largest search ever mounted in Canada, in what could be the nation's worst-ever aviation disaster. It was also the country's first high-mountain search for a large passenger plane. The odds were not good that an airliner with one dead engine experiencing icing conditions in the darkness of a fierce winter night could crash-land among the peaks and valleys of a rugged mountainous area without loss of life. The two TCA pilots were experienced flyers and the plane was sturdy, but this was extremely dangerous—in fact, near impossible—terrain in which to make a forced landing. Could any of the sixty-two people aboard still be alive? Experts thought it highly unlikely, but nobody was prepared to issue a public statement. Their job was to find the plane or its wreckage as quickly as possible. Extremely low mountain temperatures meant any survivors would be alive for only a matter of hours.

Squadron Leader George Sheahan was a veteran flyer and commanding officer of No. 121 Search and Rescue Squadron. He had rushed to headquarters as soon as he received a call about an overdue plane and immediately began making plans for the huge search that would take to the skies at dawn.

Sheahan had, like many young Canadians of this era, enlisted in the RCAF as soon as possible and from 1940 onward trained at several bases in Canada. He wanted desperately to get overseas but did not manage to do so until 1944, when he was assigned to fly Lancasters from England with the famous No. 419 "Moose" Squadron. He was cited in dispatches for his cheerfulness, good humour and "determination to attack the enemy." His confidence was contagious and a boost for his colleagues. These characteristics would serve him well in the arduous and difficult task he now faced.

He hoped desperately for a break in the weather but meteorologists were not optimistic. The storms that had battered the coast for days were still very much alive. Sheahan's rescue squadron had a vast array of planes that could be utilized for all kinds of emergencies. His group had trained hard and been involved in numerous minor searches in the rugged coastal region, but a high-mountain crash of a large airliner was outside their practical experience. He knew it would tax their knowledge and skills.

Sheahan got in touch with TCA headquarters in Montreal, and

within hours the company had called in many of its own staff to help meet the emergency. They discussed their responsibilities with Sheahan and it was decided they would take on the difficult task of informing the families of the three crew members and of the passengers that their loved ones were missing in the BC mountain wilderness.

Meanwhile, Sheahan assigned five men to begin checking maps of the area and to sift through any reports received that evening from airports or observation stations. Others soon joined them as Sunday night wore on into Monday morning. Everyone assigned to the Search and Rescue squadron was anxious to help and the building became crowded as personnel heard the news and came in early.

The RCMP and other emergency organizations throughout southern BC and northern Washington state were also alerted to the crisis. Commercial radio stations in Vancouver and across Canada got word of the incident.

The searchers hoped for a miracle and until they knew otherwise would continue to be optimistic, at least publicly. As reports of the missing plane hit the airwaves, friends, family and strangers all prayed for the passengers, hoping against hope that the crew had somehow managed a belly-landing on a mountain slope or in a valley without totally destroying the plane.

Flight Lieutenant Lin Dowsett led the group scanning detailed maps of the Cascades, looking for the most likely areas where Flight 810 could have been put down in an emergency between Hope and Vancouver. The walls of the search headquarters were soon covered with maps, charts and technical data on the aircraft involved. Emphasis was placed on the few promising areas where the North Star might be found unscathed.

Vivian Clarke and her sons Jay and Teddy were at home sleeping when a TCA pilot knocked at the door. Sitting in her living room full of the spirit of the coming Christmas season, Mrs. Clarke listened quietly as the pilot told her what was known. She said, "I was dreaming that we were both up flying. I used to be a stewardess on his flight. It was a beautiful night. Smooth . . . stars . . . snow on the mountains." She was having difficulty facing the reality of the pilot's words.

As the terrible news sunk in, her first concern was for her children. She went to tell her sons that their father's plane was missing but that there was still hope that he and the others aboard were safe. A few hours later the former nurse and TCA stewardess told a *Vancouver Sun* reporter, "Jack is a wonderful pilot, I know because I flew with him. I know he will be all right, if there was a way out." For the rest of the night she listened to radio newscasts that repeated over and over what had happened, but there were no new developments and no encouraging information for the anxious woman and her boys.

Another TCA pilot was sent to the home of co-pilot Terry Boon to break the news to his mother and stepfather, Mr. and Mrs. R.E. Marshall of North Vancouver. They too had been busy writing Christmas cards and getting ready for the holidays. Mrs. Marshall later stated, "I was never worried before because he had talked me out of it. He told me he always flew by the book." She was proud of Terry's achievements; from early childhood he had participated at every opportunity in a wide range of sports and was a keen baseball player. The Marshalls also stayed glued to the radio hoping for good news. Mrs. Marshall noted the irony of how this was to have been Terry's last flight on a North Star before he began training on the new Viscount.

The first edition of newspapers on Monday told the unfolding drama, the *Vancouver Sun* announcing in a banner headline, "Vast Hunt for Sixty-Two on Lost Plane." In vivid detail, the stories stressed the wild, rugged country in which Flight 810 had gone down with descriptions, pictures and maps of the mountains. All nearby news sources sent out reporters and cameramen to cover the event that was soon making headlines around the world.

Squadron Leader Sheahan's search armada took to the sky at first light, which on this December day was not until nearly 8 a.m. Rain and howling winds bore out the pessimistic assessments of the weather experts. Meteorologists could find nothing to indicate there would be any early break in the storms. They contacted weather stations up and down the coast and into the farthest points of Alaska, but the patterns remained unchanged. It was a far-reaching Pacific storm affecting most of the province and raging into the Prairies as it swept east.

As word of the disappearance spread, telephone lines at the Jericho search headquarters and at police and fire stations were flooded with calls. Many people reported they had heard a variety of loud noises and explosions during the night that could have been the missing North Star. Still others claimed they saw a huge flash like lightning, or strange lights falling in the mountains. Most of the calls came from around Chilliwack and Cultus Lake, some of them mentioning Silvertip Mountain or the area nearby. Although some of the messages seemed fanciful, nothing could be overlooked. Air and ground parties would investigate every lead as soon as possible.

Sheahan put together a fleet that included planes from the military, TCA and the RCMP, as well as privately owned planes. All were to take part in a massive search of a vast area encompassing more than 7,000 square kilometres. Much of it was harsh, inaccessible country stretching from Squamish in the west to Yale in the east and from Bellingham, in coastal northwestern Washington state, east to Mount Baker, the volcanic giant that dominates the skyline in this region.

Military doctors and nurses were on standby in case they were needed and parachutists were ready to board aircraft if any sign of wreckage was found. An RCAF Otter was prepared to fly in twelve members of a specially trained ground rescue team to the nearest possible landing area should Flight 810 be located in an inaccessible spot. The RCAF flew its Piasecki helicopter up the Fraser Valley to Chilliwack. It was a powerful machine capable of lifting heavy loads and had often been used for hauling small planes from crash scenes. Members of the Chilliwack Flying Club also turned out in large numbers, realizing theirs was the closest small airport to the likely crash site. Volunteer pilots in Calgary were also ready to fly in case there was any chance that the North Star had turned to the east before disappearing. Members of the local Alpine Club of Canada chapter and other mountain climbing groups had volunteered to lead a party up the mountains once the crash site was located.

Sheahan outlined three scenarios that might have overtaken the crippled aircraft. "First, the plane became lost after its radio failed; second, the pilot lost control completely; and third, the aircraft blew up." Sheahan felt the least likely scenario was a mid-air explosion. Captain Art Rankin, TCA's western superintendent, said the messages received from the plane indicated that while Clarke had shut down No. 2 engine after an instrument warning, there was nothing to suggest the engine or the plane was on fire. Unfortunately the tape recording of the conversation was not absolutely clear. The experts believed that the red light could have been an instrument warning only. The same warning light had been reported blinking on other North Stars before due to electrical malfunctions. Clarke had routinely shut down the No. 2 engine and there was no reference to fire in any subsequent messages.

Sheahan admitted he had only a vague idea of where the missing plane might be found, but he did have some clues. The RCAF now had a report from the US radar station at Birch Bay of the unusual radar blip and its disappearance. Those in charge at search headquarters had little doubt that it was Flight 810. The operator had tracked it to a point thirteen kilometres south of Princeton before it disappeared. The last sighting had been near 8,500-foot-high (2,600 m) Silvertip Mountain in the Skagit Valley. It was the only verified report indicating where the airliner could have gone down. Sheahan immediately concentrated the hunt there.

Where exactly in this white wilderness was Flight 810? Was it on Silvertip or another mountain nearby? The plane carried no emergency locator, the device mandatory on today's aircraft that sends out an emergency radio signal in the event of an accident. This tool is built to withstand terrible impacts and has meant the difference between life and death in air accidents in recent years. Aircraft in the 1950s also had no "black box," the nearly indestructible flight data recorder that provides valuable technical data to crash investigators. This airborne search team had only the human eye to scan for signs of life or plane wreckage.

"Try the peaks first and then get down to 100 to 200 feet, but for God's sake be careful," Sheahan warned his pilots, particularly concerned that some of the keen volunteer flyers might try operating beyond their capabilities. This was a tall order as severe weather conditions continued unabated, turning into an even heavier mixture of rain, sleet and snow with increasing gale force winds howling in from the Pacific Ocean.

"The extreme turbulence over the mountains is buffeting even the biggest planes, but we will stay with it as long as we can," Sheahan said during a December 10 press briefing. "The wind was very strong at the time the plane disappeared, up to 90 knots [167 km/h], but we feel his actual ground speed was only about 65 knots [120 km/h]." If anything, the search commander's description of the conditions his fleet would be encountering seemed to downplay the grimness of the situation. Sheahan may have been trying to simplify the task in order to boost enthusiasm for the difficulties ahead and keep spirits high among the searchers. He was also anxious to bolster the morale of worried-sick families across Canada who were following every news report and hoping for some encouraging words.

Reporters knew they were onto a story packed with drama and went looking for vivid accounts of the situation. In its next edition the *Vancouver Sun* described the "great fleet battling wild atmospheric conditions . . . knocking aircrew out of their seats and breaking safety equipment." Flying Officer H.S. Gamblin was navigating a DC-3 when wild winds shook the plane. He was hurled out of his seat and smashed his head against a bulkhead. One Air Force pilot was heard on the radio asking, "How are the other crews? Mine are all sick."

RCAF Flying Officer Hal Cameron, a radio operator attached to the Search and Rescue squadron, wrote a story for Canadian Press about his role as a spotter aboard a Canso aircraft. "The air was as rough as a bucking bronco," he wrote. "The squadron was keyed up when the search began. All of them were hoping to find someone alive and they considered their mission to be much more than just a job. They were dedicated to doing the best search possible." He explained how the plane dropped down to about 200 feet from the ground searching the mountainsides and valleys. "The air turbulence was terrific. We were rocked up and down so hard we had to gain altitude or be forced down into a dangerous situation. One of the crew was thrown about and hurt, but still we kept on looking." Cameron, who later became a public relations man for an airline, finished the story with the comment, "Getting into a narrow canyon, especially when there is rough air, is pure dynamite."

There were more than fifty planes involved in the initial hunt, including twelve from Vancouver and two from Comox. While the total search area was vast, the fleet concentrated on small areas, where they made rapid descents through infrequent breaks in the cloud, meaning there could be several aircraft in a fairly tight area. The dreaded possibility of a mid-air collision was ever-present.

Sheahan was concerned about the awful conditions. He frequently warned against taking unnecessary risks.

The fleet converged on an area of fifty kilometres around Hope. Pilots did their best to scour river valleys and mountain slopes when they could dodge through holes in the heavy cloud cover. While the assistance of the small planes was vital for this, Sheahan knew that wind gusts could toss them around the sky like leaves. Later in the day Sheahan reluctantly grounded all the small aircraft because the storm—and visibility—was worsening.

The proximity of the small Chilliwack airfield and the concern felt by members of the Chilliwack Flying Club for a plane lost in their backyard had brought volunteers out in large numbers. They were particularly motivated when they learned that a young teller from the local bank was on board. Donald Holden had been on his way to spend three weeks' Christmas vacation with his family in Winnipeg.

W.A. (Butch) Merrick of Abbotsford was up at first light on Monday in a privately owned Bonanza, and was followed closely by Ralph Clarke in a plane that belonged to the club. Some of the pilots were reluctant to give up, even when urged to do so by Sheahan. Club members had participated in many searches over the years but none as dramatic or as important as the hunt for Flight 810. This new endeavour easily eclipsed the club's work in the disastrous Fraser River flood of 1948, when they had delivered badly needed sandbags to crews trying to hold back the raging river. They had operated in some very tight situations, but this search called for careful systematic visual inspection of steep mountain slopes and some very tricky flying.

Reports of flashing lights on Kilgard Mountain, a small peak near Abbotsford, briefly raised hopes for the rescue teams. Constable Lorne Newsom of the Abbotsford RCMP headed a group up Kilgard's slopes in pouring rain, slipping and sliding in terribly muddy conditions. They found the light, but it was a sad disappointment. The blinking they hoped had come from survivors was actually from a microwave relay system. High winds had blown open a louvre and exposed a bright light.

One small search plane spotted another hopeful sighting: a fresh swath cut through the forest north of Harrison Lake. But it turned out to be a fairly recent logging cut.

On the ground, military personnel and police near Hope were knocking on doors to check reports from people who had heard loud noises around the time of the plane's disappearance. Some had also reported seeing strange lights moving in the mountains. One logger said he had witnessed a "blinding flash" in the vicinity of Silvertip Mountain. The mountain remained a focus of the search. On this first day, planes circled several times around the peak to check what at first looked like a slash mark in the snow. One pilot descended as close to the mark as possible until observers in the aircraft decided the slash was merely an irregular pattern carved by the high winds. Searchers swooped down to check similar patterns elsewhere, but nowhere could they find the plane. All potential traces turned out to be natural phenomena.

At search headquarters, Sheahan and his men plotted every report as it was relayed to them. Knowing how essential it was that the plane be found as soon as possible, they were anxious to avoid overlapping efforts.

The Vancouver public got many first-hand accounts of the searchers' experiences from newspaper reporters, radiomen and cameramen who went along for the ride. The three Roughriders who had stayed behind in Vancouver were up all night listening to radio reports. One of them, Martin Ruby, said, "It's an awful tragedy. We are just sitting here and hoping." Also waiting and hoping were three young women, the roommates of missing crew member Dorothy Bjornson. Olga Lychuk, Evans Smith and Theresa Hughes had all been in the same TCA stewardess class with Bjornson and all four had moved to Vancouver to take jobs only that summer.

The early darkness of a mid-winter evening ended the first day's search. Aircraft crew members flying over the area on scheduled flights kept a watchful eye for any signs of fire or flares. No one saw anything.

The federal Department of Transport immediately set up a board of inquiry that included D.D. (Desmond) Murphy, district superintendent of air regulations; W.R. Lavery, air systems inspector; W.M. Johnson, inspector of air regulations; and Joe Bertalino, aircraft inspector.

As the first day of searching ended, TCA vice-president of operations Herb Seagrim arrived in Vancouver from Montreal. With him were J.L. Rood, TCA's director of flight operations, and Jack Dyment, director of engineering. They immediately went to search headquarters to be briefed by Sheahan. If the flight had indeed gone down, this would be the airline's second major crash in less than two years.

A TCA spokesman emphasized that the North Star was well equipped with de-icing mechanisms and could handle just about any conditions. He stressed that in an emergency, the aircraft could fly with only two of its four engines in operation, but veteran flyers cautioned it could not be done for any length of time.

9

Sixty Planes on the Search

With the passing of each hour, the search for any survivors of Flight 810 became more and more urgent. On December, 11, 1956, the second full day of the most extensive air search in Canadian history, sixty planes took to the air. The weather continued to be difficult for much of the day with heavy precipitation and strong winds tossing around the smaller aircraft, but everyone was eager to help.

Twenty pilots took off at first light when it appeared there would be a break in the cloud cover, but the fast-moving system off the Pacific brought clear blue patches of sky for only a few minutes before the clouds clamped down again. The armada followed a prescribed search pattern mapped out by headquarters and included twenty RCAF planes from Sea Island, three Lancasters from Comox, three RCMP craft, two TCA DC-3s, and one DC-3 from Canadian Pacific

Airlines. There were also three helicopters from the Department of Transport and more than twenty-five light aircraft from assorted companies and flying clubs.

Sheahan repeated the instructions he had given the day before, emphasizing the need for low-level flying where possible, but stressing safety at the same time. The last thing anyone wanted was another crash or another missing plane. His warning was prophetic for Flying Officer Don Gilchrist, who had to leave the hunt early. His twin-engine Canso lost one engine, but he made it safely back to base.

The air search continued as long as possible, but by 2:30 p.m. Sheahan was again forced to ground the crews and halt the hunt. There was next to no visibility, and cloud-shrouded peaks posed a hazard for everyone. The search leader felt the frustrations of his flyers—all capable men, proud of their record in search and rescue, but unable to beat the elements. It was a tough situation. On the few occasions when they were able to get under the thick clouds and see the ground, it was like looking at a blanket of cotton wool punctuated by dark mountain peaks. The heavy snow grew deeper by the hour.

One of the many volunteer pilots who readily took to the skies was Al Stewart, a logging contractor and Chilliwack Flying Club member who lived at Harrison Hot Springs. He spent many hours during the search at the controls of his Taylorcraft float plane. Stewart knew the area well. He had bought the plane primarily for business and flew it frequently up and down Harrison Lake and Jervis Inlet to reach remote logging sites. He scoured the search area for hours at a low level, his tiny plane tossed about by the high winds, but he saw no sign of Flight 810.

Of the many reports that arrived at search headquarters from various sectors, army corporal Melford Henwood's was one of the most intriguing. Henwood explained that on Sunday night he was driving alone about eight kilometres south of Chilliwack, looking for good skiing areas. He got out of his car and was sitting on the hood making notes when he looked up to the southwest. "Suddenly I saw the most beautiful flash of light I have ever seen," he said, "a huge brilliant ball of a million different colours. It began to drop and three smaller balls of light seemed to break off from the main one." The sight was accompanied by a loud blast like a huge bomb, and he felt the explosion as he sat on his car even though he was miles away. Henwood said, "Everything seemed to tumble down the mountain," but as the lights descended, they disappeared. He reported what he had seen to the RCMP. On three subsequent weekends he made short trips into the area, convinced he would find the plane, but he was unable to pinpoint the blast's exact position.

Another compelling witness report came from George Hill in Ryder Lake. The weather on Sunday night was "dirty" with wind and a bit of snow, he said. He explained that his sister had been visiting from White Rock and had left to go home, but she later returned as a large tree was blocking the road. Hill put his power saw in his truck and headed out to clear it. He was about to crank it up when he heard a loud noise, almost like thunder. He looked toward Mount Slesse and saw a flash. Then he continued clearing the road. The following day he heard about the missing plane and told his friend Einar Erickson, who had lived in the area much longer, about the thunderous noise. Erickson talked to searchers but was told they were looking in the area closer to Hope. When they came back later to talk to Erickson, the snow on Mount Slesse had deepened considerably, and he told searchers they would have to wait until spring to see anything on the craggy peak known locally as "The Fang."

Later, a woman named Mrs. Wiebe who lived just east of Chilliwack reported that she had been out doing chores just after 7 p.m.

on December 9 when she looked up in the sky and saw "a whole basket of stars" descending to the ground.

One other eyewitness report of the crash was not made public for some time. Verny Nelson of Cultus Lake recently recalled what she saw on the night of December 9: "It was after dark and I was taking the clothes off the line. I had four babies at the time and mountains of diapers. The wind was getting up and the storm was getting worse. I was out there because I didn't want to lose the diapers. I heard a plane flying low towards Mount Slesse and then I saw a big ball of fire. I forgot about the incident. I guess I was too busy with the babies, but a year or so later I read that the crash looked like a ball of fire and it twigged my memory. It must have been Flight 810 that I saw."

Search planes carefully covered the area around Slesse when weather permitted but found no sign of wreckage. It snowed heavily in the Cascades for much of December, building up by feet each day and likely concealing any trace of the crash site. To the large number of people involved in the search, Flight 810 seemed to have vanished.

At his Vancouver home, TCA Captain Jack Wright met with several pilots and reporters to discuss the disappearance of Flight 810 and to venture guesses about how and where the plane could have gone down. Wright had been at the controls of the Super Constellation flying from Toronto to Vancouver when Flight 810 fell silent. It was shortly after 6 p.m. that Wright spoke to Clarke to warn him about severe icing conditions over the Cascades and to suggest he fly at 19,000 to 20,000 feet in order to avoid them. "He was climbing and thanked me for the information and that was all there was to it," Wright told reporters. "He gave no indication of any trouble." Wright went on: "When I was in the area the icing was not severe but that is a condition that can change very quickly. It was a little choppy, but you could not call it rough. It may have been better or worse where he was. If he had come back down this way, even if he lost radio contact, he would have been able to see the lights of the airport all right." Wright added that the North Star could be handled on three engines as easily as four. "The percentages are pretty good."

10

Hope Dwindles

The *Vancouver Sun's* headline story on December 12 stated what many were beginning to fear: "Time Running out in Search for Missing TCA Airliner." One official said searchers still held out hope of finding survivors, but that it would not last. Below-freezing mountain temperatures combined with heavy snowfall to dampen chances of survival. An impenetrable wall of cloud and fog still covered the whole Fraser Valley from Vancouver to Hope. Thick clouds were nearly solid between 800 and 13,000 feet.

Despite all the bad news, families of some passengers found hope in the story of another mountain crash in BC's Interior region six years earlier. A twin-engine Canadian Pacific DC-3 had been flying smoothly from Vancouver to Penticton, about 320 kilometres and ninety minutes from Vancouver, when passengers noticed the thick cloud cover thinning and then saw trees coming up fast below them. There was a frightening thump as the DC-3 slammed into trees, shattering the plane's cockpit. The twelve passengers aboard were thrown about the cabin but injuries were minor. Up front, however, the cockpit was crumpled. The pilot was dead and the co-pilot unconscious.

Passengers removed the co-pilot to the cabin where he was looked after by the stewardess and a nurse who happened to be on the flight. One man used a fire extinguisher to douse flames shooting from one of the engines while other passengers burned branches to alert passing aircraft.

The DC-3 had crashed on 4,000-foot (1,220 m) Okanagan Mountain, about 40 kilometres from Penticton. Searchers located it quickly. The next morning, planes dropped medical supplies and three military parachutists onto the scene before ground parties made their way to the wreck. The co-pilot died, but all twelve passengers were rescued alive.

Okanagan Mountain sat at a lower altitude and was a much gentler mountain than the rugged Cascades, but relatives of those flying aboard Flight 810 still hoped and prayed that there might be survivors this time too.

Sheahan grounded his RCAF crews until 6:30 a.m. Thursday, although four were kept on standby at the airport on the chance there might be a sudden improvement in the weather. Meteorologists continued to predict poor flying conditions, but the weather patterns in the coastal mountains could change suddenly.

In Vancouver, TCA vice-president Seagrim praised the missing Captain Clarke's flying abilities. "His performance was flawless, he made all the right decisions at the right time," said Seagrim. The company executive emphasized that Clarke was absolutely correct in deciding to return to Vancouver rather than continuing on to Calgary on three engines. He declared this meant Clarke would have had "a downhill run" into Vancouver instead having to maintain a high altitude for 480 kilometres over the treacherous Rockies. Seagrim, a long-time flyer with many hours of experience, told reporters, "He followed exactly all company and transport department regulations. He did what I would have done under the same circumstances."

Seagrim then explained TCA's educated guess as to what could have happened to Flight 810. The company's experts were convinced Clarke ran into a combination of problems including turbulence, icing and strong headwinds that suddenly, and for unknown reasons, rendered the aircraft unmanageable. "We are just as anxious as the public to establish the cause of the crash. We will do so as soon as it is humanly possible," he stressed. The TCA executive also said the airline was in the process of installing weather avoidance equipment in most of its planes, though the equipment would not be installed in the North Stars because they were scheduled to be taken out of service in two or three years. He emphasized such equipment probably would not have helped Clarke anyway.

In Montreal, TCA president Gordon McGregor, a veteran bush pilot with years of flying many types of aircraft, said, "Something other than a faulty engine caused the loss of the aircraft." He added that there was no indication that Clarke had any problems with No. 2 engine after it was shut down.

Sheahan's prayers for even an hour of good weather continued to go unanswered. Only one helicopter managed to make a survey of 3,000-foot (900 m) -high Sumas Mountain, the tallest of the lower-level peaks in the Fraser Valley. There had been several reports of sounds coming from the mountain and flashing lights, but a careful, low-level search found no signs of an aircraft accident. A ground party had already climbed to the top of Sumas and turned up nothing.

Despite that, C.F. Munro of Vancouver, whose brother John had been a passenger aboard Flight 810, desperately wanted to go up

Sumas Mountain because of persistent reports from the area. Wearing only street clothes, Munro started out on a search for his brother, accompanied by two local residents. Two newspapermen, *Sun* reporter Tom Ardies and photographer George Diack, got wind of his intentions and decided to tag along. They scrambled up through ragged bush to the 3,000-foot (900 m) summit. Although it was not a steep climb, they sometimes had to crawl on hands and knees to get over rocks, fallen trees and mud. It was wet and cold. From the top, the group scanned the horizon, but there was no trace of the missing plane.

Munro was exhausted and depressed by the time he climbed down. He felt, however, that he could not have rested until he had gone to investigate all the reported sightings and noises on Sumas for himself. "It was horrible, it was horrible," he told the RCMP of his search. The police treated him for exhaustion before he went home to Vancouver.

Ground Search

By Thursday, December 13, with the air search still hampered by bad weather, the emphasis of the hunt for Flight 810 switched to the ground. Volunteers were willing to risk life and limb as they slogged through the mountains for any trace of the aircraft. They concentrated their efforts around Silvertip Mountain and the Skagit Valley, rugged undeveloped terrain laced with logging roads and frequented by campers and fishermen. In winter, only four-wheel-drive trucks dared venture in.

A hastily assembled group of sixty-two loggers, trappers, RCMP and outdoorsmen set off from Chilliwack into the bush following old logging roads and using equipment loaned by forest companies. RCMP Staff Sergeant William Wallace led the party. In Hope, Mannix Logging contributed nine four-wheel-drive vehicles to the hunt.

The sleet, rain, cold and high winds made the going tough for the ground searchers. Even on well-defined roads, they frequently ran into tangles of old downed trees and underbrush that made progress slow. They soon realized they were engaged in what was likely a hopeless search. They had no clear idea of where Flight 810 had gone down but they pressed through the wilderness all the same.

Twenty-five planes kept up the search on Thursday, but were limited to a two-hour window early in the day before visibility again forced Sheahan to call it off. One flyer stated, "This area is known as one of the roughest pieces of real estate in BC and the weather certainly isn't helping things." Asked by a reporter if he thought they could find the North Star, he added, "I don't know. We can only hope and pray and keep on looking."

There had been additional spotters with high-powered binoculars aboard some of the larger planes but again nothing was detected on the snow-covered peaks. Flying Officer J.R. Lindeman expressed disappointment that nothing had been seen despite the fact that "a pair of trained eyes can pick out a mountain goat or a man at a range of ten miles and can tell the difference between a lantern and a campfire."

As he had done previously, Sheahan kept three aircraft circling above the cloud cover in the hope that a break would allow them to hurry down for a look. But the weather was unrelenting. All anyone could see was a heavy sheet of white.

12

Snow Obliterates Everything

On Friday, with more than five days elapsed since Flight 810's disappearance, officials finally agreed publicly that there was little chance of finding anyone alive. They believed that exposure would by now have killed anyone who remained alive after the initial impact. Several feet of new snow had fallen since Sunday night, and overnight temperatures were again well below freezing.

Grieving relatives, reluctant to give up hope, still publicly grasped at straws. One unidentified RCAF spokesman told reporters that if Clarke had been able to belly-land the aircraft—as opposed to slamming headlong into the face of a mountain—there was a chance some of the people aboard could have survived. A few relatives clung to this last hope, no matter how unrealistic it was.

There was a slight break in the weather Friday and Sheahan sent out eighteen military aircraft. They were able to get closer than ever to the peaks but spotters still found nothing. One of the six brothers of missing stewardess Dorothy Bjornson volunteered as a spotter aboard a search plane, still hoping to find a sign of his much-loved only sister. Hope was now all but gone, and searchers acknowledged that the week's snowfall had likely covered any trace of Flight 810 and the sixty-two people on board.

The search continued Saturday, six days after the accident, but at a scaled-back pace. It was now a recovery mission. A few new developments came to naught. Loggers at the base of Silvertip thought they heard voices on the mountain but could find no one; an avalanche spotted on Lady Peak raised interest but there was no plane; and an oil slick on Chilliwack Lake was investigated. Samples taken from the slick were rushed to BC Research Council for identification, but tests found they were not aircraft fuel. Reports continued to keep search headquarters busy and on their toes but they led nowhere.

TCA's McGregor issued a statement to relatives stating that because of the elapsed time, "hope was diminishing" for finding survivors. This had been the unofficial view almost from the outset, but it had not been stated publicly in deference to passengers' relatives. Families were also forced to acknowledge the loss when Queen Elizabeth issued a news release, stating, "I have heard with deep concern of the tragic loss of the Canadian airliner in the Rocky Mountains. Please convey my sincere sympathy to all the relatives of those who have lost their lives." The message presupposed that all had been lost, and like many in the United Kingdom, whoever wrote the message thought the Rocky Mountains swept all the way from Alberta to Pacific coast tidewater.

Another TCA official felt it necessary to say that the company had given no thought to grounding the company's remaining North Stars, although there had been no cries from the public or from members of the press that this should be done. In the mass of reports pumped out about Flight 810—some of them totally erroneous and speculative—there was no criticism of the aircraft's history or performance, or of the capabilities of the two men in the cockpit.

On Tuesday, the eighth day of the search, the weather suddenly improved. Breaks in the clouds, however, only revealed the extent to which new snowfall would hinder the search. Snow now enveloped the Cascades, putting an impenetrable blanket over any wreckage that might be strewn on the heights. Sheahan had twenty-eight aircraft on the lookout, but the exhaustive hunt was taking its toll. Some of the crews were starting to feel the fatigue caused by hours of tricky flying and even longer hours of waiting for favourable conditions. Sheahan was not ready to give up the search and with improved flying conditions the next day, he again scrambled together between thirty and fifty aircraft during the short daylight hours.

TCA agreed to allow two long-time captains, Art Rankin and J.D. Storie, to partially re-enact the events of December 9. They tried to replicate the path of Flight 810—from Clarke's first radio message about engine trouble to the plane's disappearance. The pilots took off from Vancouver in a North Star and followed the route as closely as possible, making the southwesterly turn at Hope and trying to hold the same altitude, speed and rate of descent that the missing crew had radioed. They also manoeuvered the North Star in a manner they thought would simulate the effects of the headwinds and strong gusting side winds that had been reported at the time. Sheahan and his staff had pored over maps and provided all the information they could in order to make the simulation as accurate as possible. In the end, the simulation provided no new clues and was a considerable disappointment to searchers who had hoped they would at the very least be able to spot some sign of the aircraft on the mountains near Silvertip.

The very deep snow cover had by then changed the contours of all the mountain rises and valleys. It was many feet deep and had buried everything on the ground, making it impossible for spotters to see anything no matter how low they flew. By Thursday, December 20, the search was at a standstill. Severe weather grounded all aircraft. Only one plane was cleared to circle the area, but it did so in virtually zero visibility.

TCA and airport officials held their breaths briefly when another North Star incident occurred only eight minutes out of Vancouver Airport. Once again the pilot reported engine failure. The fate of Flight 810 loomed large in everyone's mind while they waited for the plane to return, but this time the North Star, with twenty-nine passengers aboard, landed without incident.

Two weeks after Flight 810 disappeared, it became generally accepted that all those aboard had perished. There was, however, one report that stirred renewed interest and brought the airmen very close to discovering the crash site. It came from keen amateur pilot Butch Merrick, the Chilliwack Flying Club member and Abbotsford café owner who had taken to the air in his Bonanza on December 10 and continued the search at every opportunity. On a December 23 journey, he and a passenger both thought they saw something glittering through the clouds high up on the side of Mount Slesse, the last mountain in the Cascades before the Fraser Valley opens out like a fan leading to the Pacific Ocean. Merrick descended within sixty metres of the slope. "There was absolutely no doubt in my mind," said the pilot, explaining what he had seen. He described a piece of a fuselage and part of a propeller. His passenger, Paul Martin, corroborated the assessment. Merrick radioed Bellingham, the nearest air station, which contacted Vancouver air traffic control. Two other Chilliwack Flying Club members searching that day reported a similar glittering on Slesse.

The next day, December 24, TCA sent out a Viscount piloted by

Captain Art Rankin with eight special observers aboard to check Mount Slesse with high-powered binoculars. Unfortunately, they were a day late. There had been more heavy snow overnight, and the TCA crew saw no evidence of wreckage, just the same vast white expanse that greeted them everywhere.

Despite his insistence that the plane lay high on the slopes of Slesse, Merrick was told that what he and the other small plane had seen was probably just a log, even though Merrick pointed out that the area in question was well above the tree line. Whatever the reason, both TCA and rescue headquarters failed to follow through on this lead and so the focus of the search moved elsewhere. For some inexplicable reason, the sightings by the Chilliwack flyers did not make the major newspapers at the time. It was not until May 23, 1957, that the *Sun*'s "flying photographer" Bill Dennett finally detailed Butch Merrick's account.

Christmas 1956 was a sad, quiet day for the families and friends of those aboard Flight 810. Little searching was done over the holiday, although all military and civilian aircraft flying on regular flights over the Cascades kept special watch. Nothing new was seen or reported. As Thursday, December 27 dawned and people of the Lower Mainland returned to their jobs, an unnamed Department of Transport official was quoted in the *Vancouver Sun* stating that a navigation error might have been to blame for the crash. The official noted that Flight 810 had strayed well south of the Green 1 corridor it should have been taking back to Vancouver Airport. This was the first time anyone had suggested the possibility of such an error. TCA president McGregor refrained from comment, instead issuing a statement that he had been holding back until after Christmas. It was an official acknowledgement that all aboard Flight 810 had lost their lives. McGregor emphasized that efforts to locate the aircraft and recover the bodies would continue as soon as weather conditions permitted. The search director, Squadron Leader Sheahan, also spoke out about the situation. "There is no longer any hope of survivors as far as we are concerned and so the sense of urgency in the search has now gone," he said. "We don't intend to do any further searching in weather that can endanger air crews."

Mountain experts said there would be little hope of sighting wreckage until after the spring melt, although they pointed out that some of the highest parts of the Cascades were never bare. The scarred mountain faces contained holes and crevasses big enough to swallow up several North Stars. Some experienced climbers even voiced the opinion that without a lucky break, it was possible that the wreckage might never be found.

Search headquarters spent a good deal of time during this period plowing through a stack of new messages ranging from the rational to the bizarre. One recommended that searchers try a simple form of divining rod; another said to talk to people who had seen the downed plane in their dreams. It took up valuable time, but every letter and call was checked out just in case there was something that might help.

On December 27, the *Vancouver Sun* ran a story that criticized the lack of planning for ground searches, maintaining the job could be handled better by the RCMP or the Canadian Army than the Royal Canadian Air Force. It touched off an ages-old inter-service rivalry.

Mel Cassman, a volunteer searcher from Vancouver, told the *Sun* he was amazed that soldiers based in the area had not been called in at the outset to help with the ground search. He said the Army could have supplied everything from trucks and snowmobiles to radios and emergency rations instead of crews having to scrounge everything they needed from local companies and agencies. He pointed out that the searchers on the ground had no thermal sleeping bags or other heavy-duty outdoor equipment. *Sun* reporter Phil Hunt wrote that the ground search he had covered was chaos, and that the RCAF officer in charge had no knowledge of the terrain or local conditions and no experience running such an operation.

Local RCAF spokesman Sergeant Buzz Sawyer, a popular figure with the media and later a TCA employee, came to his service's defence. He contended that in such a huge and rugged area, and with no clues as to where Flight 810 might have gone down, even a large troop would have had little chance of finding anything. Sawyer pointed out that there were only a half-dozen trails in the area between Chilliwack and Hope, and that it was very dangerous country in winter.

Squadron Leader Sheahan, who had left Jericho headquarters for the valley when ground searches began, said he had asked the Chilliwack army base for equipment and the detachment had responded immediately. A colonel at the base, Arthur Fraser, said the demands had been minor. He had been given clearance to provide as much equipment, including snowmobiles, as was requested and could have assigned twenty men trained in mountain climbing and winter survival to the initial operation. But the requests never came.

The RCMP stayed well clear of the issue and the Air Force/Army squabble. The *Sun* claimed the search system needed an overhaul, even though a high-altitude, large airplane crash had never happened before in Canada and there was little experience in tackling this kind of emergency. The paper pointed out, however, that in Washington state and elsewhere in the US, emergency planning was generally more thorough. It was stressed that Washington's Mountain Rescue Council had very clearly defined and delegated responsibilities.

Despite the Flight 810 disaster, 1956 had been a solid year for TCA. McGregor released a year-end statement showing that the air industry was booming, particularly in North America. The International Air Transport Association (IATA) reported that airlines had carried 90 million people, up from 78 million in 1955, and were predicting 100 million for the coming year. McGregor's passenger line had flown one billion miles in Canada and the US and 188 million overseas. Parcel express, mail and cargo figures also were up sharply. Air routes around the world were expanding. Scandinavian Airlines announced that in February it would start flying over the North Pole on a run from Copenhagen to Tokyo. It would make the flight of 12,600 kilometres in thirty hours, compared to fifty hours for standard routes through the Middle East.

IATA also announced that air fatalities had dropped in 1956; 0.62 fatalities for every 100 million passenger miles, compared with 0.76 in 1955.

Over the course of the winter, TCA officials wondered if they would continue to buck the fatality trend when a Super Constellation with eight crew and sixty-one passengers reported trouble over the mid-Atlantic. The pilot radioed that a starboard motor had quit. He requested permission from Shannon Airport in Ireland, which was controlling this leg of the flight, to descend from 23,000 feet to 11,000 feet in order to reduce the possibility of icing. Permission was given, but the pilot's problems compounded when a second engine cut out about 1,120 kilometres west of Shannon. Fortunately for all concerned the pilot radioed again to say the engine had restarted. He decided to continue on to London where the aircraft landed safely.

Also over the winter, the RCAF announced it would be replacing all North Stars with the bigger, British-made, turboprop Britannia, which would provide more speed, range and capacity.

The extreme conditions of a very tough winter continued to sweep through the coastal mountains of BC, where the wreckage of Flight 810 and the bodies of the sixty-two people lay somewhere in the deep snow. In January, TCA hired a Vancouver company to make an inch-by-inch aerial survey of the 800-square-kilometre search area, but the work done by Aero Surveys Limited failed to locate the North Star. The search could not resume fully until well into the spring, when the snow would begin to melt and streams and rivers would fill with runoff from the mountains.

13

Crash Site Discovered

For all the time and money spent on the massive air fleet and well-provisioned ground crews, it would be an insurance adjuster with a deep passion for climbing who finally cracked the mystery of Flight 810. Elfrida Pigou was a mountaineer. The mighty peaks of the province thrilled her. To tread where few had been, to gaze upon a seldom-seen horizon—these were the joys that gave meaning to her life. Every spare moment was spent planning the next trek or reliving a difficult climb. Nothing was as important to her as the mountains, and she continually searched out new uncharted slopes that would challenge her skill and stamina.

Pigou was quiet and unassuming by nature, unless she was in the mountains. In her mid-forties, she was unmarried and worked as a claims adjuster in North Vancouver. Among local alpine clubs,

For Elfrida Pigou, the magnificence of Slesse's peaks and the possibility that a climbing party might find the missing plane along the way made an irresistible challenge. The Province.

she was admired as a distinguished mountaineer, one of the best in the country, a wiry little woman who could edge her way up a sheer rock face or trek tirelessly for miles up steep slopes. Club members often traded tales about the huge packs she strapped on her back and the way she was undaunted by the unpredictable difficulties and setbacks of a new climb. She lived by the same credo as George Mallory, the famous English climber who, when asked why he wanted to risk life trying to scale Mount Everest, replied "Because it's there."

Pigou had a desire to be first to the top of numerous untried BC peaks. She was a member of BC climbing parties that made first ascents of Mount Raleigh, Mount Gilbert, the Cleaver Peaks, Homathko Peak and Mount Essex—all towering members of the Coast Range north of Vancouver. She also made several first ascents up Mount Mummery and found new routes up Snowpatch and Bugaboo spires in the Purcells. The Coast Range, and the Cascade Mountains in particular, were her playground. She may have been the best woman mountaineer in North America at the time.

Pigou had begun hiking as a girl, climbing the hills around her home in Vernon. She entered the University of British Columbia on a scholarship and graduated in 1931. She and her father moved to North Vancouver and she took a job in the insurance industry, developing outside interests in opera, plays, art exhibits and lectures.

In 1948, however, her life changed. She attended a climbing school in the Garibaldi region of southwest BC, located slightly southwest of the now famous Whistler–Blackcomb skiing area. Her keenness and natural ability quickly resulted in invitations to join some special expeditions. She was encouraged to attend an alpine camp at Maligne Lake, where she took on some tough climbs in the Rockies and qualified to become a member of the Alpine Club of Canada.

Back in BC she joined a group preparing to climb Mount Gilbert, near Bute Inlet on the southern coast. In 1952 it was the closest peak to Vancouver of over 10,000 feet (3,048 m) that had not yet been climbed. The expedition was defeated by bad weather, but the group successfully returned two years later. In an article documenting the expedition, Pigou explained her love of the sport:

> It's a different world up there, so very clean, the contrasts so sharp, the beauty so immense. In the cities things clash . . . I think clashes bother people even when they are not entirely aware of them. But on the mountains there is no clash. The climbing itself is all rhythm and harmony, and wherever one looks there is also rhythm and harmony as though it had been designed by some very great artist.
>
> There is a contrast—a spray of wildflowers can be silhouetted against a snowy jagged peak in the mountains. In the city one can be very artificial, almost everything is artificial and it doesn't show so much. But it is impossible to be on a mountain climb with others, in danger sometimes and often uncomfortable, without eventually discovering the real people underneath. For instance I defy anybody to be artificial very long while 50,000 mosquitoes are biting.

She was well aware of the dangers inherent in mountaineering. On several occasions she had been rescued by fellow climbers and often volunteered when others were lost or in trouble. She was among a group publicly recognized midway through 1956 for finding one of three missing hikers.

By this time her reputation was legendary. Pigou had scaled some sixty mountains, including most of BC's tallest. She was always prepared to look for a missing climber, but like fellow Vancouver mountaineer Waldemar "Fips" Broda, she did so under one very clear condition: "I believe that so long as there is a chance to save a life, even the faintest chance, a climber should take the risks involved. I don't believe it should be done for the dead."

Pigou sometimes told stories about her adventures to fellow climbers around the campfire. Once, while she climbed a glacier with a small party, she missed her footing and fell into a deep crevasse in the ice. As she was falling, her ice axe dug in and stopped her plunge. She dangled there, hanging onto the axe until her companions were able to haul her back up to safety. Even in crisis Pigou took time to enjoy the view from the inside of the glacier. "It was like being in a glass-encased box," she would say.

If climbing was her passion, photography was her hobby, and she was good at it. In the summer of 1956, several of her shots were exhibited in a show of mountain photography at the Vancouver Art Gallery. The emotion she felt for the mountains came through in her photos and she won an award for one of her pictures.

That same summer, four months before TCA Flight 810 went missing, Pigou was a member of a group that attempted to climb Mount Slesse. But the peak was shrouded in cloud, and they never reached the top.

After the disappearance of Flight 810, the Alpine Club's Mountain Rescue Group made a number of forays into the snow-clad mountains following up on tips from various sources. The group formed a committee made up of *Province* reporter Paddy Sherman, Austrian alpine expert Fips Broda, club president Roy Mason and Pigou. They worked closely with Sheahan. As winter turned to spring, Sherman made a number of flights into the mountains to check on receding snowlines. He was eager to be among the first to spy the plane and break the story.

True to her character, Pigou began planning a second assault on Mount Slesse in early 1957. She had become convinced it was the most likely site of the crash. The sheer magnificence of Slesse's peaks made it one of her favourites and now there was a chance that a climbing party might find the missing plane along the way. It was an irresistible challenge.

As the early days of May arrived, the spring melt began in earnest. Pigou and two fellow climbers planned a journey up Slesse that would begin just days before a concerted, highly organized search for the wreckage was to be launched by a new search and rescue team. Sheahan was ready to assign several helicopters to remote ridges where, armed with powerful binoculars, observers would carefully sweep the peaks and slopes looking for where the sixty-two people had died.

At about this time, a climbing fatality shook the local mountaineering community. Stan Johnson, thirty-one, of Vancouver—a recent arrival from England—fell to his death while descending a glacier in the North Cascade Mountains of Washington state. This accident prompted Sheahan to warn his observers and climbers to be extra careful. The emotional stress of Johnson's death might have delayed or discouraged some from facing the dangerous conditions of a spring melt, but Pigou was undaunted. She had studied reports about the crash and felt there was a good chance the plane was on Mount Slesse. She was also anxious to succeed where she had failed the previous summer to see "The Fang" in all its glory.

Weather conditions for the area were good, although highly unpredictable as always in the Cascades. Pigou started out from Vancouver on Saturday, May 11, with Geoffrey Walker, a twenty-eight-year-old immigrant draftsman from England, and David Cathcart, a twenty-three-year-old civil engineer who had emigrated from Northern Ireland just six weeks earlier. They were all experienced mountaineers but Pigou was clearly the leader; she was the most experienced and most familiar with the area.

The three drove east through the Fraser Valley toward Cultus Lake and Vedder Crossing before taking a road along the Chilliwack River until they were as close as they could get to the mountain by road. They parked the car and began the climb, reaching 1,900 feet (580 m)—an eight-kilometre hike into the forested lower levels of the mountain—before camping the night in two rundown shacks.

The following day they set out at 6 a.m. facing a tough 6,000-foot (1,830 m) ascent. They also had to be careful to leave time for the return descent before darkness fell. It was rough country. Thick forests and fallen logs soon gave way to loose soil and rocks where it was difficult to find a secure footing. As they edged their way higher, they met melting, slippery snow and ice. Gradually, it became less treacherous as they reached higher elevations and firm frozen ground. Pressing on, they reached 4,600 feet (1,400 m), where they were confronted with a new threat. Mist surrounded them, visibility became more difficult and they found themselves in thick cloud. As the trio moved higher, the snow became deeper, and their laboured progress slowed again. The atmosphere was muted, murky and moisture-laden, and the sodden climbers feared they would have no chance at all to see the triple spires of Slesse.

Geoff Walker later described it as "a difficult climb with a lot of rock work and some snow climbing." Visibility was so bad they became lost on the scarred western face of the mountain. Gullies many metres deep surrounded them on all sides.

Pigou was searching for the same gully that she had worked her way up the year before. Time was of the essence. She had to

Mount Slesse was only the third climb David Cathcart had made in Canada—an expedition to remember.

decide whether to continue on and try sighting the wreckage of Flight 810 through a shroud of mist or minimize the risks and start back down the mountain. She decided to forge ahead toward the gully, but she took a wrong turn. It was a fateful mistake.

As Pigou later wrote in her diary, "Curiously, Lady Luck took a hand. We turned up the wrong gully. It was pretty well snow all the way. We got to the ridge top. Dave [Cathcart] wasn't feeling so good and he decided to rest there. It was pretty cold and Geoff [Walker] and myself carried on up the ridge towards the peak."

Pigou soon realized her mistake. The gully she had chosen was farther from the peak than she had been on the earlier expedition, but it was too late to turn back, so she and Walker carried on. It was by then close to 1 p.m. They had been climbing for more than six tiring hours.

As they trudged upward, Walker spotted a piece of paper lying on the snow. He picked it up and was surprised to find it was a map of Sydney Airport in distant Nova Scotia. Upon closer inspection he found that it was a TCA map detailing the approaches to the runways and the radio beams in operation. Its significance didn't immediately register with the two climbers. Walker stated, "We didn't think much of it at the time as it seemed quite dry and perfectly clear. It could have been dropped quite recently, we thought."

As they plodded toward one of the pinnacles that crowns Mount Slesse, Pigou spotted a large object that seemed out of place in the glistening snow. She climbed over to investigate and found a twisted piece of aluminum about two feet long. It lay ten feet below a small ridge. It had rivets and some metal attachments, and looked like part of a plane. Suddenly, the scene all made sense. "As soon as I saw the plane fragment I thought of the missing TCA North Star. There was no doubt what it was," she said.

Pigou and Walker became the first to know exactly where Flight 810 had hurtled through the sky and crashed. They stared in awe at the ragged rock high above them. As if in recognition of what they had found, the cloud lifted and they were able to gaze at the peaks for the first time. Pigou later wrote, "It looked as if the aircraft had been flying a few feet higher, it would have made it." In the following days, many were to make the same assessment: if only the plane had been flying just a little higher. One hundred feet might have made the difference between life and death for all on that stormy night.

Excited by their discovery, Pigou and Walker looked for more clues in the immediate vicinity. They found three smaller pieces of metal. One was a pressure fitting of some sort. There was also another twisted piece of aluminum and a hexagonal piece of metal that looked like blue anodized aluminum. "The sheet metal looked as though it might have been shattered by a terrific explosion," Walker stated.

As they hunted further, the pair found three bits of shattered plywood and some sewing thread wrapped around a piece of paper. They also picked up several pieces of newspaper and gathered the detritus together into a small rock cairn to mark the location for future search parties.

Strangely, it seemed there were no signs of any human remains. They carried on up the slope, but the mist was again descending and in the increasing gloom they saw only sheer drops to ledges below and the stark rock face of another pinnacle. Where was the remainder of the wreckage? From what they were able to determine from their position on this west slope, Pigou and Walker concluded that the bulk of the North Star's shattered frame had plunged more than 2,000 feet (600 m) down a sheer face on Slesse's east side and now lay far below them, at some distance from where the plane had initially struck the mountain.

Knowing they had no more time to explore, Pigou and Walker headed down taking turns carrying the large piece of metal. More than two hours had elapsed since they had left Cathcart and headed out on their own. Alone in the cold and still ailing, Cathcart was beginning to worry about his climbing companions and how he would make it down the mountain if they failed to return. The weather seemed to be worsening and he did not know this mountain. It was only his third climb in Canada. His apprehensions soon vanished as he saw his companions emerging from the gloom, excitedly calling his name and waving a large piece of shiny metal.

The three huddled around their intriguing find to examine it more carefully. They were able to see that it bore the initials TCA and a series of numbers. It was the perfect piece to identify the aircraft but there was a difference of opinion as to what to do with their precious discovery. The two men wanted to write down the information and leave the jagged scrap on the mountain, but Pigou wanted to cart it out and, as leader of the group, she won the argument. She elected to carry it herself, strapping it atop her pack for the descent. After all the stories and rumours, she felt that others might not believe their tale unless they had proof that the wreckage of Flight 810 had been found.

Wet and cold, the climbers were exhausted by the time they reached the cabins where they had spent the previous night. They ate sparingly and then began the eight-kilometre bush hike back to the car.

It was early Monday morning, about 2 a.m., when they finally reached Vancouver. Each went home and caught some much-needed sleep before they informed anyone about their find. Rising only a few hours later, Pigou was uncertain of what to do, but it seemed only natural to get in touch with her friend and fellow Alpine Club member Paddy Sherman. She reached him at the *Province* newsroom and related the dramatic news. He was as excited as she. For him it would be one of the most important stories of his career and he was on it in a minute. He offered to drive to Pigou's home in order to pick her up along with her precious piece of cargo.

The discovery of the wreckage of Flight 810 triggered a story that would dominate the news for weeks to come. It also brought Pigou, a very private woman who would normally have shunned publicity, more attention than she had ever wanted.

14

Paddy Sherman's Scoop

Paddy Sherman paused for only a second as he hung up the phone. Then he turned to city editor Bruce Larsen and yelled "North Star!"

It was May 13, 1957, five months since the plane had vanished. The *Vancouver Province* newsroom staff sprang to life. Larsen and managing editor Bill Forst took a quick look around to see who was available and after a hurried conference began to assign coverage of the story. They were determined to hit the streets with an extra edition in order to scoop the rival *Vancouver Sun* before staff there got wind of the discovery. There was a lot to do and they knew that every minute the story was delayed gave the competition across the street more time to discover what Pigou and her companions had found the previous day. It would soon be the talk of the mountaineering community and then word would spread like wildfire.

Rival companies then owned the *Province* and the *Sun,* and competition between the two news staffs was fierce. The *Province* was one of Vancouver's first dailies, and in 1957 it prided itself on being a solid family newspaper. But its circulation was still suffering from a recent newspaper strike that had lasted three years. In that time, the *Province's* circulation dropped to a distant second behind the *Sun's,* although it had managed to hang on to a core of faithful subscribers.

The *Sun* was a less conservative paper, brassier and flashier than its main competitor, which some disparagingly referred to as the "Old Lady of Victory Square." But the *Province* got a big break when it broke the story of Flight 810's discovery.

When Sherman and Pigou arrived in the newsroom, Sherman called TCA's regional manager Norman Donnelly and gave him the identification numbers on the piece of wreckage so that they could be confirmed as belonging to Flight 810. Photographs were taken featuring the tiny woman holding the piece of the plane's wing. The shots would be readied for the front page of the special edition while Sherman and Pigou set off for Donnelly's office.

Veteran *Province* columnist Jean Howarth described in gloating terms how her paper had managed what she termed "one of the big scoops in Western Canadian newspaper history." Howarth wrote,

> Yesterday was that kind of day at the *Province.* While our competition and every news service on the continent were frantically searching for them, we had the three mountain climbers who found the TCA wreckage in our library . . . Engravers, printers and pressmen were alerted. Aircraft were chartered and reporters and photographers were on their way to key points in a matter of minutes. Just to give you an idea of how fast the people who put out newspapers can work when the emergency hits I'd like to tell you that

our engravers produced the exclusive *Province* page one picture of the plane wreckage in fifteen minutes. Normal time is anything from half to an hour or more.

Howarth explained how three reporters went separately to bring the climbers to the office, despite their assertions they should be at work. Others headed for the Vancouver Airport, Chilliwack and Search and Rescue headquarters. One reporter, Bob Reguly, who used to be a forest fire smoke jumper, took off in a small plane with the intention of parachuting into the site. Reguly would have needed his head read if he had actually jumped into the mountain wilderness.

Howarth then presented the paper's moral dilemma: should they turn the wreckage over to authorities, "or should they hang onto everything for late editions"? They decided on the former. The pieces were delivered to TCA officials and identified immediately as belonging to Flight 810. Howarth plumbed hypocrisy with a sentence that stated, "We wiped the tears away with our exclusive extra which hit the streets at 10:30 a.m. No opposition newspaper came out until two hours later." She said keeping the climbers away from the competition was all very "cloak-and-dagger" and described how they were smuggled out of the *Province*'s office and into a hotel until the afternoon. "By that time our home edition, with the whole story, was on its way to your house. And you read it only in the *Province*."

When the "extra" hit the streets, the editors of the *Sun* grabbed copies and pinched much of the material, doing a fast paste-up job using stock pictures of mountains and the North Star so they could put the story on the front page of their own home editions. If the *Province* had held off until the home edition, the *Sun* would have missed the story completely. Still, *Province* street sales that day were far above normal.

Frustrated *Sun* editors made huffy charges that the *Province* held back the release of information from the authorities in order to get the story out first. It may have been Howarth's story, however, that prompted *Sun* reporter Jack Brooks to come up with a feature slamming the *Province*. A hard-nosed reporter, Brooks quoted an unnamed spokesman as saying officials were "fighting mad" at the newspaper staff. Brooks wrote, "TCA officials have asked for a full report about withholding wreckage evidence and failure to report the sighting of the wreck to them or to the police."

Sun publisher Don Cromie became so annoyed with the *Province*'s inside information that he had the question raised in the House of Commons. The reply from the Minister of Transport was that Sherman's access came from his being a member of the Mountain Rescue Group, not from his being a journalist.

Every available reporter in the *Province* newsroom began digging into an aspect of the biggest story since the plane's disappearance. The librarian dug out all the clippings on the disaster from December and January and all the pictures on file were taken out and reviewed. For the days following the discovery of the twisted wreckage, Mount Slesse was the scene of the action. Through its climbing reporter Sherman, the *Province* kept its edge in the newspaper coverage.

Many were concerned that victims' families had received news of the plane's discovery only by hearing radio reporters read the *Province* extra edition on air. Officials had had no time to inform relatives that the crash site had been found.

Sherman was determined to see the site of the disaster for himself as soon as possible. His help would be needed to guide search parties up the precarious slopes and he wanted to be sure he was ready. It would keep him on top of the news. His instincts as a reporter told him this would be one of the biggest stories of his life. With his experience and contacts in the mountaineering world, he knew he had an opportunity to lead coverage of one of the major stories of the year.

Sherman was then a twenty-eight-year-old reporter who had arrived in Canada with his wife, Maureen, in the summer of 1952 from England, where he had worked on several daily papers. At this time there was a significant shortage of reporters on both Vancouver papers, which were happy to hire the experienced journalists emigrating from the UK. Paddy picked up work almost immediately. At the *Province* he quickly established himself as one of the paper's top reporters. And he had done it despite heavy competition from other experienced British reporters who had flocked to Canada after World War II.

Sherman was a dark-haired, wiry young man—quick-moving, quick-witted and destined to go far in the Canadian newspaper business. He never missed a chance to seize a good story. He was a newsman first even though his career took him into management. Many years later, in 1986, he would become president of the Southam Newspaper Group, publisher of both the *Ottawa Citizen* and the *Vancouver Province,* the paper where he got his Canadian start. Of that appointment he recently commented, "I said no to the Southam job three times before I finally agreed to it."

At the same time Sherman became a well-known mountaineer involved in climbing the magnificent local mountains that had first influenced him and his wife to move to Vancouver. He had organized an expedition in 1954 that made the first ascent of Mount Gilbert, and among the group was an avid, much-admired mountaineer named Elfrida Pigou. They went on several other ascents together, including the Homathko Icefield and Mount Queen Bess.

When Sherman and Pigou arrived at the Department of Transport office, it was already crowded with expectant officials who, upon examining the evidence, quickly agreed the missing plane had been found. The serial numbers on the wing fragment confirmed it belonged to the ill-fated TCA North Star. The sight of the ragged metal lying on a table revived all the feelings of frustration that had been so much a part of that futile mid-winter search as well as the poignant memories that plagued those who knew the pilots and passengers.

With the evidence he needed in his hands, Norman Donnelly relayed news of the discovery to Search and Rescue headquarters at Jericho Beach. Once again Sheahan went into action. Men, equipment and planes were readied and a contingent quickly set out up the Fraser Valley. The search team knew there were two immediate priorities: the recovery of bodies from the site and a detailed investigation of the wreckage in an effort to uncover any clues to what brought down the North Star.

From the *Province* newsroom, Bruce Larsen sent the first reporters and photographers on their way up the valley to Chilliwack where they hoped to charter an aircraft to the mountain.

TCA's Donnelly put Sherman on notice that they would need his climbing expertise to help in assessing the problems of a recovery mission on Slesse. And, just as he had hoped, it put him in a prime position to cover the story. The operation would require the skills of many mountaineers and Sherman would select and lead the most experienced to the site, giving the *Province* the inside track as the drama unfolded.

15

The Stark Impact Scene

About an hour after the *Province*'s special edition hit the streets with a headline blazing the news that the wreckage of Flight 810 had been found, Paddy Sherman was in a small helicopter on a reconnaissance mission to Mount Slesse. The mountain's three sharp peaks were still partially wrapped in cloud but visible through frequent breaks.

Veteran flyer Evan Bullock of Okanagan Helicopters was at the controls of a tiny Bell G-45, a glass-enclosed bubble that danced on the winds swirling around the ominous summit. As the two men peered down, Bullock guided the helicopter lower and lower until they could actually see tracks left by Pigou and her companion two days earlier.

Bullock used all his skill to touch the machine down several times on the slopes of the mountain. At times, he brought the rotor

Province *journalist and mountaineer Paddy Sherman photographed the wreckage of Flight 810 dangling from the face of the precipice of Mount Slesse. Photographs courtesy Paddy Sherman.*

blades spinning mere feet away from the granite mountain walls. A sudden gust of wind could have meant serious trouble.

Near the top of one peak they were able to observe the ledge where Pigou had found the shard of aluminum wreckage. It was a rocky outcrop jutting about thirty feet out from the face of the summit. Sherman wrote, "From our airborne perch, we could see the serenity of the little hamlets in the Fraser Valley. But the spot, so near where sixty-two people died, was a place of stark and savage wilderness . . . A hundred feet to the right and the North Star might have gone through the gap in the ridge. A hundred feet above and it would have cleared the terrible teeth of the mountain." Sherman believed, however, that another mountain ridge farther along and lower still could have claimed the plane if it was rapidly losing altitude.

Sherman and Bullock returned to Chilliwack to report what they had seen to a growing group of officials who were developing recovery plans. They knew that whatever attempts they made would be difficult and dangerous. RCMP officers from Chilliwack were dispatched to the crash scene, and Sheahan's ground forces were almost ready to set up base camps for those who would try to retrieve bodies.

As Pigou's group had found earlier, dense forest made progress by foot very difficult. Sheahan decided it might be necessary to clear a landing site for helicopters if no natural clearing could be found. The recovery mission was fraught with obstacles. Flying in and out was going to be difficult at any level because of the still-unpredictable weather conditions. At the same time, a vast and growing pack of reporters and cameramen from media outlets across the country was pressing leaders of the expedition for the latest news and pictures from the mountain.

The *Sun*, desperate to catch up on the story, chartered a plane and sent veteran reporter and aviation specialist Ron Thornber to describe the scene. In colourful style, he wrote, "I flew where death wears a snaggle-toothed grimace high atop Mount Slesse almost within sight of the broad and peaceful farmlands of the upper Fraser Valley."

Thornber then added, "It is a dangerous mountain even to fly near . . . I've never seen a more rugged or crueller peak than Slesse in many flying hours on air searches over coastal and interior mountains of BC."

Public interest in the discovery intensified when the federal Department of Transport's Desmond Murphy revealed that one of the victims might have been carrying a money belt stuffed with $80,000 in cash, the equivalent of around $600,000 today. Murphy said the revelation came to him from a cousin of passenger Kwan Song, who was returning to New York from Hong Kong and was known to have been carrying this amount when he left the US. An airline official said he hoped the disclosure wouldn't start a stampede of would-be searchers "going into the hills" looking for the money. The thought may have crossed many minds but most realized that the remains of the aircraft lay so high in the mountains that only those with superior climbing skills and endurance would be able to reach it. It was no place for treasure-hunting amateurs. Nevertheless, in the weeks to come, several would-be grave robbers would be turned away. They failed to realize that with wreckage spread over thousands of feet, the chances of finding the booty on the snowy, crevice-laced slopes were almost nil.

Plans were made following Sherman and Bullock's helicopter survey to launch another climbing expedition, this time to view the area some 2,000 feet (600 m) below the peak where they believed most of the wreckage lay. Although her first climb had been exhausting, the indefatigable Elfrida Pigou prepared to scale Slesse again Tuesday, less than forty-eight hours after her previous attempt, in order to lead a team from the Mountain Rescue Group to the site.

They began by flying to a rough base camp at 2,500 feet (760 m). This time Pigou was with Sherman, experienced local climber Jack Russell and Fips Broda, considered among the best mountaineers on the West Coast. They would climb the west side of the mountain where Pigou had made the initial discovery. Another experienced group consisting of Roy Mason, Ian Kay, Joseph Hutton and Bussell Yard would tackle the mountain from the east side, looking for the main section of the plane and any human remains.

Fips Broda was by far the most experienced climber in the group. In 1954, along with John Dudra, he had made the first known winter ascent of Mount Slesse. A thirty-four-year-old Austrian immigrant, Broda had been a wartime instructor for his country's mountain troops, training them on some of the most difficult mountains in the Alps. He was tough and demanding. His personal achievements included scaling Mont Blanc—Western Europe's highest mountain—and other daunting peaks in Austria, Germany and Italy. He had become a member of the Alpine Club upon arriving in Canada in the early '50s and had quickly become one of BC's best climbers.

Sherman, Pigou, Russell and Broda began the steep ascent of the west face soon after touching down at base camp. As they climbed, several large boulders were dislodged and clattered away down the slope below them. Sherman accidentally loosened a rock that spun toward Broda, narrowly missing his head. Even a glancing blow could have swept him off the ridge. Pigou also had a narrow escape when a boulder Sherman and Broda inadvertently freed sent some large stones hurtling in her direction. Fortunately Pigou heard them coming and ducked behind an overhang as the small slide thundered past. After what seemed like an agonizing length of time, they heard her call out that she was not injured.

As the four climbers approached the wreckage, they saw an amazing scene. Some of the North Star's control cables had wrapped around an outcrop near the peak. A piece of the aircraft's nose was tangled in the wires, prevented from falling down the precipice to where the remainder of the plane had come to rest. The tangle contained part of a man's body still clad in what appeared to be a TCA uniform. The whole mess dangled over the sheer face of the peak—surreal and horrifying, it was a sight that held the climbers spellbound.

After careful inspection, the group decided unanimously that it would be impossible to recover any of it. The force of the impact had carried this section of the plane to the west side of the pinnacle while the remainder of the aircraft had plunged 2,000 feet (600 m) to a basin below.

Sherman later wrote, "We examined this piece of wreckage carefully, crawling between it and the rock face. We felt it might be technically possible to remove it, but very hazardous, as any interference would probably dislodge the entire mass. Just a few feet below the outcrop, the steep slope quickened and curved out of sight, dropping vertically for more than 2,000 feet (600 m) to the snowfield below. Anybody trying to retrieve this fragment of a body might easily wind up there too."

The climbers took some spectacular pictures of the wreckage hanging from the control cables and of Slesse's foreboding rocky face. While her companions surveyed the area below on the east side of the peak, Pigou stopped, repelled by the thought of what they might find. "You men were all in the war," she said. "You are more accustomed to this sort of thing than I am. I don't want to go any farther."

Carefully, the male team members picked their way down through difficult terrain, deciding not to rope together because the area was covered with broken rocks and shale. The surface was extremely unstable; if they were tied together, one slip could drag them all down.

The broken remains of two engines were discovered, more

evidence of the tremendous force of the impact. The men continued the search, finding an array of personal belongings. There was a little girl's dress; a woman's half-burned shoe; a pair of swimming trunks; some new earrings still attached to the display card, maybe meant as a gift for someone; a few coins; a half-burned newspaper; some mail; a purse; and tangles of twisted metal. Some of the material they found was difficult to identify because it had been shredded by the impact and then singed as the North Star's loaded gas tanks erupted in a brilliant flash visible for miles around. The three men steeled themselves as they inspected the melting snow, finding first a hand, then an arm and other unrecognizable human remains. As they searched the impact area, planes and helicopters whirled around Slesse's peak carrying reporters and cameramen whose film and pictures would show viewers and newspaper readers the awful destruction at the impact point.

Sherman and his companions could see the four other mountaineers well below them, working their way up the mountain. The lower party shouted and waved, their voices carrying faintly but clearly to those above. They explained their progress and what they had discovered. The area where the main wreckage lay was about half a kilometre long by some 90 to 180 metres wide. For the most part it was steep terrain covered with snow and ice and strewn with large rocks. This steep bowl, or cirque, carved out by glaciers thousands of years before was poised at the edge of a precipice with a sheer drop of another 1,000 feet (300 m) to a level farther below.

The snow was melting rapidly in the warm sunshine of May, and the mountaineers faced ever more slippery conditions with an increasing danger of avalanches. There were fears that the roar of so many airplane engines might trigger a slide, but fortunately they disturbed only the quiet of the mountain, not the surface of the snow pack.

In the group climbing the east face, Roy Mason said that there were crevasses in Slesse's scarred surface that ranged from three to twenty metres in depth, and many of them were packed with snow. In some ravines the climbers could see clothing, luggage and aircraft parts. Mason saw no trace of human remains. There were several small avalanches while the group searched. The unstable bowl where much of the plane's remains lay was an extremely dangerous area, racked constantly by snow and rock slides that increased in intensity during the afternoon hours.

Mason estimated it would be late summer before the snow would recede enough to allow a full recovery effort. He said the area was like a glacier and that as time went by, constant ice and snow movement would gradually bury much of the wreckage. There was also a possibility that the entire ice pack would shift enough to push the whole mass 1,000 feet (300 m) down to the next ridge without warning. The entire basin was extremely treacherous and the party remained at the site for only about two and a half hours.

The *Vancouver Sun* knew it was being badly beaten on the story of the year. After the descriptions of the first climb made the pages of the *Province*, the *Sun* sought out Fips Broda and on Wednesday, May 15 carried a headline story containing his account of the climb as told to reporter Simma Holt.

The Broda story began, "Not one of the sixty-two persons aboard the Trans-Canada airliner on December 9 suffered even a split second of agony when the plane crashed. I stood beside the wreckage of that crashed North Star and I can tell the families of those aboard the plane that nobody lived beyond the moment of impact. I am sure from what I saw at noon Tuesday at the top of Mount Slesse that the aircraft exploded the second it struck. All those inside died instantly."

The story emphasized how agonizingly close Flight 810 had come to clearing the spiked peaks of Slesse. "If the plane had gone thirty feet to the right or about forty or fifty feet higher the sixty-two people aboard could be alive today," Broda maintained.

There were variations in witnesses' guesses as to how near the plane came to missing the granite face at the top of the mountain, but the estimates were all within a hundred feet of each other, and all emphasized how close the roaring aircraft, fighting to maintain altitude, had come to skimming the top instead of colliding with the peak. Beyond was the relative safety of the Fraser Valley flatlands.

Broda went out of his way to chastize those who thought they might stroll up to the scene and search for the money reported to be carried by the Chinese passenger from New York. "I know that any inexperienced climber, any money-hungry treasure hunter seeking the reported $80,000 in the wreckage is doomed to die," he said in the story. "One wrong step and he would slide 2,000 or 3,000 feet down the cliff to death. It is a dangerous place and the rocks have been loosened by the crash." Broda didn't mention it, but in all probability the passenger would have spent most of the money because he was at the end of his trip.

Broda stressed exactly how difficult the climb was. He said that immediately after landing his team had faced a 1,400-foot (435 m) snow climb at a thirty- to forty-degree incline and then encountered rock at even steeper slopes. "At some points we climbed at a seventy-five-degree angle," he stated.

"We had to cross narrow shelves, drop and scale sharp cliffs all over loosened rocks with a sheer drop off below us of something like 2,000 to 3,000 feet," he said. He was of the opinion that it would be impossible to escort officials or technical inspectors to the mountaintop unless they were experienced climbers. He outlined the difficulties they would have to face. "In order to bring officials of TCA or the Department of Transport in here, we would have to have seven or eight fixed ropes 100 feet (30 m) long to get them from the 5,600-foot (1,700 m) level, where a helicopter could land, to the 7,600-foot (2,300 m) peak where the wreckage is. Even then, if they are not skilled mountaineers, we would have to practically bundle them up the ropes to get them safely in and out."

Broda described how at 11:45 a.m. the team of climbers arrived at the grim site where Flight 810 had met its end. He said he found bits of wreckage, some coins, shoes, red-flowered swimming trunks and "a few letters, one addressed to one of the Chinese men on the plane and another to a trust company." He added that they saw part of a hand, a foot and other parts of bodies amid the main wreckage. Broda said they had been told not to touch anything and to replace anything they had to move as nearly as possible to where they found it.

Emphasizing that the force of the impact must have been unbelievably massive to break up so much rock, Broda told Simma Holt, "It is a terrible scene up there. I saw a lot of death in the war, but this is something I will never forget."

Broda's story shared the front page of the *Sun* with coverage of another airplane incident, this one involving popular federal cabinet minister and MP Jimmy Sinclair, father of Margaret Sinclair (who would marry Prime Minister Pierre Elliot Trudeau). On May 14, Jimmy Sinclair was flying to Whitehorse from an election campaign meeting in Dawson City when a Yukon snowstorm struck quite suddenly. Pilot Roy Connelly immediately decided to put his Beaver aircraft down on partially iced-over Lake Laberge. The plane skidded to a halt through the slush without damage. Connelly and Sinclair made their way to shore over thin ice and then walked three kilometres to Whitehorse. Sinclair, a former member of the Air Force, said he agreed with bush pilot Connelly's quick decision, quoting the old service phrase that "there are old pilots and bold pilots but few old, bold pilots."

16

McDonald's Inquest

Vancouver coroner Glen McDonald, an active, gregarious man who didn't mind seeing his name in print, was a popular figure with reporters. He had been assigned by BC Attorney General Robert Bonner to head an inquest into the disaster. McDonald rushed to Chilliwack accompanied by his morgue attendant, David Quigley, who was carrying a body bag when the two men first met with reporters. McDonald helped to bring into focus the awful tragedy aboard Flight 810. He told the horde of newsmen that his aim was to get "one single identifiable body" from the wreckage so that he could attribute a legal cause of death for all those on board. He stressed that he had "not yet been ordered to retrieve the remains at all costs, but the coroner's job is to recover the bodies of all those who come to unnatural deaths."

The coroner was taken on a helicopter over the main wreckage at

Paddy Sherman (left) and world-class Austrian climber Fips Broda on Mount Baker in the 1990s. Photograph courtesy Paddy Sherman.

5,600 feet (1,700 m) shortly after he arrived but no attempt was made to land. Following his aerial inspection he opened an inquest hearing in Chilliwack on May 16, 1957. The coroner stressed that while the aircraft had been identified, they had still not officially identified any one of the sixty-two people aboard.

At the hearing, McDonald asked Sheahan about the possibility of recovering bodies. Sheahan said it was impossible at the present time: "We have no alternative but to wait until sometime in July."

"Can you deny or affirm rumours of bodies in mounds covered by snow, and can they be retrieved?" asked McDonald.

"I deny the rumour," Sheahan replied.

At one point Sheahan raised the ire of the entire mountaineering community by suggesting the skills of local people might be inadequate for the task at hand. Coroner McDonald had cited a crash in a Colorado canyon for which European climbers were brought in. "This is a multi-million dollar operation," he stated, "and there is no sense relying on local options when we can get the best available within forty-eight hours."

The McDonald balloon was quickly brought down by Fips Broda, the world-class Austrian climber, who said that BC's top climbers were a match for any on the planet when it came to rescue work. He also reiterated his warning to any fortune hunters thinking of tackling Slesse: "They're doomed. They would tumble to their deaths from the many cliffs that block the path to Slesse's summit," Broda predicted.

McDonald wondered if a large group going in with shovels to the lower level could remove snow covering some of the bodies. Paddy Sherman responded, "We could remove the snow but I wouldn't use anybody but trained mountaineers. It's too dangerous." Sherman also said he doubted that any of the body parts the climbers had seen could be identified. His assertion was backed by Broda, who said the remains they saw lay in small pieces. McDonald could be a hard man to convince, but he finally deferred to the views of the mountaineers, particularly Broda on the question of outside help. Broda had, after all, been the first to climb Slesse in winter. Broda concluded his comments at the inquest by asserting the unwritten law of the Alps: life should never be risked trying to recover bodies.

The RCAF's Sergeant Buzz Sawyer confirmed just how dangerous the area was. He had been flown over the main wreck along with Quigley, the coroner's assistant. He said the mass of the North Star was perched in a precarious position. "If any attempt is made to work there," he told the inquest, "it would bring the whole mass crashing down." While Sawyer and Quigley had been hovering over the wreck, a ten-metre vertical wall of snow had come thundering down without any warning whatsoever. The whole area was so dangerous that it should be placed out of bounds to everybody, he said. Quigley supported Sawyer's views, adding that when the wall of snow collapsed, "we had to move fast to get out of there."

McDonald was still concerned with bringing out a body to establish a legal cause of death. At the close of the night's proceedings, it was agreed that he would join a party that would go in by chopper to survey the main wreckage the next day. The group was led by Sherman and Mason and included TCA vice-president Seagrim; J.T. (Jack) Dyment, TCA chief engineer; Charles Johnstone, TCA airways engineer; and Desmond Murphy of the Department of Transport, who had done much test flying of the North Star and was heading up a federal inquiry into the crash. For much of the rest of the spring and summer, though, dangerous conditions would thwart McDonald in his attempts to scour the main crash site.

A Hair-Raising Climb

Despite repeated warnings of the dangers on the slopes and the heavy penalties that could be imposed on lawbreakers, many rushed to the scene the weekend after the discovery. RCMP guards turned away more than two hundred people who claimed to be there out of general curiosity. One car full of men claimed they were just fishermen hiking in to a favourite hole, but when police asked them to open their trunk, they found an array of climbing gear. After a strict warning the "fishermen" beat a hasty retreat back down the road.

A pilot aboard a low-flying RCAF helicopter spotted two men trudging through the snow, obviously trying to reach the base camp. They had a long way to go before nightfall. Even from the air it was obvious they were entirely ill-equipped for their trek through sticky, wet snow. The pilot reported the pair. Another chopper was dispatched to bring them out. When it landed, the two didn't want to get in, so the no-nonsense pilot barked, "Get aboard, you awful little men!" They were found to be totally unprepared for the wilderness. They carried a ten-foot piece of rope and a bag of sandwiches. What they intended to do was anybody's guess. Totally unrepentant, they later tried to sell some pictures to the media, but there were no takers.

The Department of Transport's chief inquirer Desmond Murphy felt it necessary to have at least one technical expert view the remains of the North Star at its point of impact. It was a difficult and possibly dangerous trek, even for someone trained in mountaineering. Murphy, who had never scaled a mountain in his life, nominated himself for the task.

Sherman and Mason took it upon themselves to prepare the slight but strong Murphy. To improve his physical condition, Murphy hiked a five-hundred-metre track up the North Shore's Grouse Mountain every morning. It was a path adjacent to what is known today as the Grouse Grind, a tough climb made annually by thousands of people. It wasn't real mountaineering but it was still good training for Murphy's assault on the Slesse summit. After his early morning workout, he would put in a full day at his office. He had been in reasonable physical shape at the outset, and these strenuous workouts added to his muscle mass and agility, both of which would be needed for his upcoming climb.

By June 25, Sherman and Broda both felt Murphy was fit for the task ahead and the three of them were helicoptered to the Slesse slopes. Flight Lieutenant Pat Matthews, an expert at handling his craft, put one wheel down on a steep slope while the trio jumped out onto the snow and started the long hike to the top. Sherman and Broda helped Murphy as much as possible using ropes and ice axes. Broda kicked out steps in the snow to make it easier for Murphy, who later commented, "It was somewhat hair-raising to be up on that mountain." Finally they reached the impact scene.

Murphy, no newcomer to the devastation of air crashes, was appalled by what he saw. It was immediately evident to him that the North Star had been travelling much faster than anyone had thought possible, considering its dead motor and iced wings. Clarke's last messages had indicated he was experiencing difficulty maintaining altitude and figures showed the plane had lost some 6,000 feet over the course of thirty-two kilometres, a rather fast rate of descent. Still, Murphy had not anticipated the devastation he now witnessed up close. He carefully checked over propeller parts and other pieces of the aircraft that remained on the ledge. He had just an hour or two to make his observations, but it was sufficient to reach conclusions that would be valuable contributions to the official inquiry.

The three men wasted no time descending to the helicopter pickup point. Clouds were forming and visibility worsened as they returned to find there was no helicopter in sight. They had been on the slopes for seven hours and were starting to think they might have to spend the night on Slesse in a tent they had brought along for just such a scenario. In addition they had food and other supplies necessary for any unscheduled stay on the mountain. But soon, despite the deteriorating conditions, the steady beat of rotors filled the air and the RCAF chopper took them aboard.

Murphy was not yet through with his investigation. He insisted on checking the lower site where the main body of wreckage lay. But the mountain would not yet allow it. Over late spring and summer, snow conditions at the lower site remained unstable. An RCAF sergeant who visited the site on May 15 in an attempt to retrieve a body was nearly hit by an avalanche. In June, Paddy Sherman wrote that looking down from 7,600 feet (2,300 m), the 5,600-foot (1,700 m) cirque where much of the wreckage sat could be described as "disappearing—the snow is very badly broken up and there are crevasses all over and in constant motion." By late August another investigator wrote, "The snow above the wreckage still constitutes a hazard . . . I would not recommend a party working in that vicinity." Another described an avalanche of some fifty or more tons of snow that scattered in spherical shapes below him for a quarter of a mile down the slopes as he struggled to climb upward. An RCMP officer wrote, "We were out there only five minutes when two large avalanches went right over where we were working."

After many attempts, one climber stated, "The wreckage buried in the snow remaining could not be examined except at some distance due to the highly dangerous instability of the snow masses and the difficulty in negotiating the snow and the crevices between masses." In the two years between 1957 and 1959, many investigative teams climbed the mountain in search of more evidence and during that time snow at the foot of the precipice never melted.

18

"We Commit Their Bodies"

I t was late August before the snow conditions were judged safe
to permit Desmond Murphy, Coroner McDonald and other investigators to fly in to inspect the lower level. They went in by
helicopter on August 27 and 28. It was summer, but the sunshine
was interrupted several times by fast-moving squalls that turned the
slopes of Slesse suddenly bleak. High winds made close approaches
by helicopter just as tricky as they had been during the winter. The
constant background noise made by restless moving snow packs
and roaring rock falls seemed to amplify the terrible devastation
remaining from the crash. Mountaineers moved carefully among the
investigators, constantly advising them on the safest way to reach
what might be an important fragment.

Coroner McDonald and the rest of the investigators were awed
by their close-up view of the fractured North Star, shattered into

thousands of tiny pieces. The largest section was only about 4.5 metres long. The high-speed collision followed by the explosion and the downward plunge had left remains that were almost unidentifiable. Still, the experts were able to pick out some parts of the aircraft among the debris. McDonald told reporters that the tremendous force of impact had compressed one of the huge Rolls-Royce engines—originally about 2.5 metres long and 1 metre in diameter—into a chunk of twisted metal about 60 centimetres round.

While small sums of money were found, there was no sign of the $80,000. If Kwan Song had been wearing his money belt, it was probably consumed in the fiery explosion. The searchers were reminded of the football stars who had been on board when they discovered film clips of the All-Star Game lying in the rubble.

The official investigating group undertook the grim task of collecting body parts scattered among the debris. When he flew in, McDonald had taken with him as much personal information on the passengers as could be gleaned from relatives. He had descriptions of the clothes the victims had been wearing, dental records and details about surgical scars and incisions. But the bodies found were so mangled that the information was of little use. After spending two days at the crash site, the group had recovered parts of seventeen bodies, none of them identifiable. A single grave was dug—it didn't have to be very large—and the remains were buried together. With DNA testing still decades away, no passenger was ever positively identified among the wreckage.

Believing there would be a need for some form of religious observance, Coroner McDonald had come prepared to provide a simple solemn ceremony. In his pocket was a copy of the Church of England's *Book of Common Prayer*. With the weather worsening, he knew there wasn't much time. But having come so far, he wanted to do the right thing. After the remains were buried, McDonald, a former naval officer, adapted a version of the prayer used at the burial of the dead at sea. It starts, "We therefore commit his body to the deep . . ." As the small party gathered around, McDonald quietly read the prayer, substituting the word "mountain" for "deep." McDonald later recalled the moments of silence punctuated by the distant crashing of avalanches. RCMP Corporal Tom Anderson stepped up to say some words in memory of Roman Catholics who had died. McDonald and Anderson then gathered up some wood and stone near the site and built a simple cairn topped with a cross tied together with string.

The coroner knew that the victims' families had desperately wanted the bodies recovered and brought out, but he also knew the dangers involved. Any further recovery would be too hazardous. With reluctance, relatives agreed it would be fitting that all remains be buried at the scene.

A month after McDonald's trip, RCMP Staff Sergeant William Wallace led a ten-man team that was flown to the site for another search. They located parts of nine more bodies. These were also buried on the mountain, but apart from the others. Even on this occasion, Mount Slesse was unforgiving. Wallace said they had left the gravesite for only five minutes when two avalanches swept down and buried the area they had been searching under many metres of snow.

The coroner concluded that those aboard the North Star had died an accidental death, but because no identifiable bodies were found, an exact cause was never attributed. Relatives would have to wait for the Department of Transport inquiry to wrap up for answers. In view of the calamitous state of the crash site, however, it was questionable whether even the closest examination of the wreckage would provide any kind of definitive answer.

Years later, McDonald wrote in his autobiography, *How Come I'm Dead?*: "I don't think the cross lasted even a day the way the snows were coming down but there was nothing more we could do.

It was a gesture, at least." He was mistaken. The hastily built memorial he and Corporal Anderson built was still standing on Slesse more than thirty-five years later, defying wind, weather and avalanches.

"The memories stay with me vividly," McDonald would later write. "The scene of absolute, total destruction and desolation just won't go away. It proved to me that the might of nature is such that it can erase all the damage and insults man can do; that these things nature constructed and maintained must, in the long run, prevail and ultimately destroy the human and the inventions he makes in the name of progress."

Because of worldwide interest in the startling discovery of the North Star, the inquest recommended action be taken to keep the cu-

There was no sign of the $80,000 rumoured to be aboard Flight 810 and the half-mile area surrounding the crash site was closed to the general public in order to discourage treasure-seekers. George Bell photograph.

rious away from the crash site. In Victoria, Attorney General Bonner said the provincial government would prepare an order-in-council under the Land Act that would close Mount Slesse's crash site to the general public. The RCMP erected a check point at the end of the road leading to Slesse. The area was also ordered closed to all aircraft, and landing on the mountain was forbidden. There would be heavy penalties for venturing within a half-mile of the scene. "Interfering with the wreck would bring even heavier penalties, a $5,000 fine or a year in jail, and maybe both," Bonner said.

When his inquest was over, McDonald got in touch with Attorney General Bonner in Victoria and suggested that the mountain be declared a monument to the victims, something he had already proposed to TCA officials. The coroner said the Minister of Forests could declare it a provincial cemetery and impose the necessary conditions, but it would be a long time before this came to be.

"The spot where death came to sixty-two persons on December 9 is a place of stark and savage wilderness," Paddy Sherman wrote in the *Province*. "The plane's victims will have a finer monument than man could ever design, safe from predators for all time."

19

The Inquiry Board Report

Every air crash in Canada is probed to the fullest extent possible. Experts study all angles to determine how and why it happened. They carefully scrutinize possible flaws in the aircraft so that safeguards can be introduced. Investigators check the training, experience and performance of the pilots. The plane's maintenance records are studied in detail. Examiners read and re-read weather reports. Radio messages are analyzed at length and the status of navigational aids is carefully checked. Any and all potential witnesses are interviewed.

Today's investigators have many advantages over those who studied the crash scene on Mount Slesse in 1957. The two black boxes now carried on all large aircraft record how the many components of the aircraft were operating up to the moment of impact and conversations between the pilots. In 1957 this technology had not yet been developed and consequently the investigation into Flight 810 was limited.

Prior to the 1985 bombing of Air India Flight 182 over the Atlantic, the investigation into the Mount Slesse crash may have been the most difficult in Canada's history. The North Star was shattered by the initial impact and explosion, and was further destroyed when the bulk of the wreck plunged 2,000 feet (600 m) down to a lower ledge on Slesse. It then lay under a blanket of snow for five months, subject to the whims of a restless, violent mountain until it was finally found. The location of the crash made it all but impossible to retrieve much debris.

At the head of the federal board of inquiry into the crash of Flight 810 was Desmond Murphy, an investigator so focused on scrutinizing the North Star's final moments that he had risked his life struggling up Mount Slesse's slopes. His tenacity could not silence the many critics of Canadian air crash investigations, however. At this time the Department of Transport wrote the rules, inspected and licensed aircraft and ensured recommendations were implemented. It was also responsible for investigations when something went wrong.

In the 1950s, the Canadian aviation industry was relatively small at the executive level. Personal relationships inevitably formed between those involved in probing accidents and those running airlines. Many senior airline officials in both government and industry had flown together in the RCAF during the war. This familiarity led to questions from those who doubted that any board of inquiry put in place by the Department of Transport could conduct a truly impartial investigation. There was never a direct suggestion that these relationships had produced any improprieties, but following the highly visible case of Flight 810 and other plane crashes came a growing public call for greater distance between department investigators and the industry. The independent Canadian Aviation Safety Board was instituted in 1984; in 1990, it was replaced by the Transportation Safety Board, appointed by the government and reporting directly to government. Its sole duty would be to investigate all marine, pipeline, rail and air accidents under federal jurisdiction.

In 1957, the inquiry board was a three-man team set up by the federal Department of Transport that would undertake the complex and exhaustive probe into the Mount Slesse crash. The investigators—Desmond Murphy, W.R. Lavery and W.M. Johnson—knew the crew was well qualified and had been fully briefed before they climbed aboard a sound aircraft with an impeccable record of maintenance. There was no breach of any regulations and the plane's record over the years was excellent. The board knew that Flight 810 had been flying about twenty kilometres south of the southern boundary of Green 1—but why? How did they stray that far? If the inquiry board could answer that one question, much of the mystery would be solved.

The inquiry dealt first with the qualifications of the crew. The final accident report stated both pilots had received their original training with the Air Force. They were fully certified and up-to-date on checkups. In the days before the crash, the numbers of the pilots' flying hours were within regulations. At age thirty-five, and with more than 12,300 hours of flying experience, Jack Clarke had been in a position to go on to fly DC-8 jets. Captaincy of a Viscount was in the future of Terry Boon, who had been looking forward to training on that aircraft. Young Dorothy Bjornson was only a few months into her posting as a stewardess, a position offering an adventurous life for the times. Despite her relative inexperience, she was a graduate nurse and an enthusiastic flyer, agreeing to take the shift aboard Flight 810 in place of an ill co-worker.

Investigators then turned to the plane itself. The airframe had been built at the Douglas plant in Chicago in 1947, which made it relatively young. Unlike some of the airframes built at the time, the DC-3 body used in the North Star was built to last. (There are still DC-3s flying today, some seventy years after the model was introduced.) The North Star had flown more than 27,295 hours—393 hours since its last major check and just 24:24 since a minor one. Its gross weight for the flight was 76,821 pounds (34,845 k), almost 3,000 pounds (1,360 k) below the maximum permissible for take off. Flight 810 had been heavy leaving Vancouver, carrying more fuel than usual because of possible bad weather over Alberta, which may have ruled out landing in Calgary or nearby Edmonton. The alternative airports—Regina, Saskatoon and Winnipeg—would have stretched the trip for a few hundred more kilometres.

Inspectors looked at the maintenance records of the four Rolls-Royce Merlin engines and deemed them to be in good working condition. Whether the No. 2 engine actually caught fire or the warning light was just malfunctioning could not be established.

The pilots should have been able to calculate their position based on radio signals from eight nearby navigation beacons. Both the mountaintop radio transmitters and the plane's own radio receivers were the most modern available at the time. There were no irregularities with any of them the night of the crash. However, it was found that bad atmospheric conditions could have interfered with radio communications between Flight 810 and Vancouver air traffic control. The radio log showed that at least two messages could not be fully understood, but neither seemed to be of vital importance.

The inquiry board's report covered every detail starting with the arrival of the crew at Vancouver Airport and the briefing given to the two pilots, right down to the red pencil used on the charts to track the storm and dangerous icing conditions prevailing throughout the west. The investigators traced the atmospheric low that caused the storm from its centre in the vicinity of Sandspit on the Queen Charlotte Islands, moving southeastward over the BC mountains, into the

Rockies and then on to the prairies and the Calgary–Lethbridge area of Alberta. The North Star took off at 6:10 p.m., climbed out over the Gulf of Georgia, and then flew about fifteen kilometres south of Vancouver before turning east toward the Fraser Valley, the Cascade Mountains and Calgary.

Board members were given access to the record of the flight and the radio messages among Flight 810, Vancouver air traffic control and the pilots of TCA flights 7 and 4. The final report gave an account of nearly every minute of the flight.

18:10 p.m. Flight 810 departed Vancouver International Airport.

18:13 p.m. The flight was cleared to cross Westham Island at the mouth of the Fraser River, southbound at 3,000 feet or below, then to shuttle south of Westham to 10,000 feet and to request further clearance. Flight 810 and Flight 7 exchanged information about weather conditions. Then Flight 810 requested clearance to 19,000 feet from Westham Island through to Calgary. ATC (air traffic control) advised Flight 810 to request further clearance when by Westham at 10,000 feet and informed the pilot that 19,000 feet should be available.

18:15 p.m. ATC asked Flight 810 if it was by Westham yet. Captain Clarke answered, "Yes, forty-five seconds ago."

18:21 p.m. ATC cleared Flight 810 to the Calgary range as per the planned route, and to maintain 19,000 feet. Clarke did not echo this clearance, but requested departure via Mud Bay, Abbotsford and Cultus Lake. This route was approved by

ATC and Clarke was requested to report back at Mud Bay.

18:30 p.m. Flight 810 reported its estimated arrival time at Princeton at 18:58 flying at 19,000 feet.

18:35 p.m. Flight 810 reported leaving 13,000 feet at Abbotsford and five minutes later reported leaving 15,000 feet at Cultus Lake.

18:43 p.m. Flight 810 reported leaving 17,000 feet and that icing was noticeable at 16,000 feet.

18:44 p.m. Flight 810 reported encountering light to moderate turbulence at 16,000 feet and above.

18:46 p.m. Clarke radioed ATC about occasional jolts and assumed that he had gone through some cumulus tops between 16,000 and 18,000 feet.

18:48 p.m. Clarke requested permission to climb to 21,000 feet.

18:49 p.m. ATC cleared the flight to maintain 21,000 feet and to report leaving 20,000 feet. Clarke acknowledged the clearance.

18:52 p.m. Flight 810 reported that at 19,500 feet it had had a fire in No. 2 engine and that Clarke and Boone had shut the engine down. Clarke stated, "Looks like we had a fire." The report presumed that the fire was extinguished immediately as Clarke did not refer to it again.

18:53 p.m. Flight 810 reported to Flight 4, the eastbound TCA Viscount, that 810 was returning to Vancouver southwestbound. Clarke also reported to ATC that he had lost an engine and was maintaining 19,000 feet. He then reported he was losing altitude quickly and requested immediate clearance to descend.

ATC cleared the North Star to 14,000 feet on Green 1.

18:57 p.m. Flight 810 reported being on Green 1, though radar indicated he was well south of Green 1.

19:01 p.m. ATC asked Flight 810 if it would be able to hold 14,000 feet. Clarke replied that he thought so. Flight 4 then advised Flight 810 that conditions were poor at the lower altitudes and he was encountering strong subsidence and moderate ice at 11,000 to 13,000 feet. Flight 4 recommended that Flight 810 maintain altitude. Flight 810 replied that he was nearly down to 15,000 feet and would maintain it. The tone of Clarke's voice implied there would be no difficulty in doing so. ATC cleared 810 to maintain 14,000 feet or above. Flight 810 estimated that the plane would be at Hope in approximately five minutes.

19:08 p.m. Just before passing Hope, Flight 810 passed an ATC clearance on to Flight 4, which had just reported passing Princeton. Flight 810 relayed the message due to poor communication on VHF radio between Flight 4 and ATC. After this there was an exchange of messages between the two flights.

19:09 p.m. Flight 810 mentioned more precipitation. The report states that Clarke said he "was unsure as to whether it looked too good or not."

19:10 p.m. Flight 810 stated it was by Hope and requested permission to descend to 10,000 feet. ATC cleared it to cross Vancouver at 8,000 feet or above and to stay on ATC frequency.

"The acknowledgement of this clearance was the last transmission from Flight 810," the report states. The final radar plot observed at 19:11 on the military network by a US serviceman at a base in Washington state south of Vancouver indicated that Flight 810 was twenty kilometres south of the southern boundary of the Green 1 approach to Vancouver and thirty-four kilometres southeast of the Hope Beacon.

The report noted, "The radio logs of TCA and ATC do not indicate that Flight 810 was in any serious difficulty. This is confirmed by Captain Rickard [the pilot aboard Flight 4] wherein he states definitely that in communication with Flight 810 that the last contacts gave the impression that the situation was well in hand."

The board called six TCA employees and one Department of Transport official to testify. Captain J.A. (Jack) Wright, Clarke's friend who had piloted the incoming Flight 7 from Toronto, explained his last radio conversation with Clarke and the advice he had given him about icing. Captain C.L. Rickard and First Officer D.F. Moir, pilots of the eastbound Flight 4, also detailed their radio exchanges. Also among the witnesses were TCA radio operator H.E. Deeks, telecommunications supervisor F.K. Anderson, dispatcher G.W. Inglis and air traffic controller C.M. Lenaghan.

Neither the witnesses who heard the cockpit conversations live while working at Vancouver air traffic control that night nor those who studied the recordings during the investigation could detect any tension in the voice from the cockpit. There was no indication of any major emergency beyond loss of one engine. The pilot's reactions were those of a competent, experienced airman. Both Clarke and Boon were well trained in onboard emergencies and knew well the capabilities of the usually reliable North Star.

The report also noted, "Nothing in the examination of the wreckage indicated that a fire in No. 2 engine broke out again and further endangered the flight. There was no evidence at the site of any real conflagration after the impact. The only burns noted were in a patch of moss or lichen that appeared to be scorched where the North Star first struck and exploded on the mountain face. Fire could have been expected there. There also were scorch marks on a woman's shoe found amid the clothing, blankets, broken baggage, cabin material and pieces of metal scattered about or embedded in the rock by the terrible force of the crash. The main wreckage at the lower level was not charred as it would have been had a fire erupted when the plane finally came to rest."

Ground crews turned up the remains of only two engines. One was jammed into a cleft in the rock face and the other was littered in small pieces over much of the crash site. The engine in the rock face was positioned so that it could not be recovered. Relatively un-damaged propeller blades were discovered high on the steep face. The report noted that some broken pieces of the propellers had large notches in their leading edges, indicating that they were still rotat-ing under power at the precise moment they slammed into Slesse. The remaining two engines were never found.

The inquiry examined all weather reports the crew was given that night. The conditions at the time of the crash were very close to forecasts. There had been a solid cloud layer extending to 20,000 feet with embedded cumulus down to about 16,000. Flight 810 had reported severe jolts thought to be caused by the tops of clouds between 16,000 and 18,000 feet and "really good jolts" at 19,000. There was also severe turbulence below 10,000 feet. As he headed back to Vancouver, Clarke had faced moderate icing and headwinds of 120 to 160 kilometres per hour. It had been a rough ride.

Searchers had been seriously shaken by what they saw on the mountain. The report stated, "The human remains found were in terribly broken condition, none of them having all their components still attached and even torsos broken into pieces." The impact had torn seats from their moorings and hurled passengers forward.

The inquiry concluded that, considering the headwinds and icing, Flight 810 had been flying at a much faster ground speed than was originally thought. The report stated, "The condition of the wreckage supports the conclusion reached after the examination of the upper wreckage that the aircraft hit the peak at a high rate of speed, not necessarily in a vertical dive, although it probably was losing altitude rapidly. Sufficient parts of the main components were found and identified to indicate that the airplane was probably whole at the moment of impact and operating on three engines."

Relating to the speed of the aircraft, the report also explained, "It is considered the estimated normal three-engine-level flight air speed of the DC-4M2 [the specific North Star model] aircraft minus an estimated 85-knot [157 km/h] headwind would give an approximate ground speed of 100 knots [185 km/h]. If the aircraft, as is suspected, was carrying ice, this speed would normally be less. The average ground speed of the radar plots from the time the aircraft commenced its return until it disappeared was 102 knots [189 km/h]."

A lack of conclusive evidence stopped the board short of attributing any sure cause to the crash. "The possibility of further malfunction or failure still exists . . . and it is unfortunate that the difficult terrain, poor weather during the summer months and wide scattering and inaccessibility of the debris prevented a complete examination to determine this point conclusively."

Of the dozens of safety belts that investigators found, only one was actually still buckled. They concluded that all the others had burst open on impact.'

The board reasoned that whatever happened aboard the plane caught the passengers and crew by complete surprise. "It is reasoned," the report went on, "that whatever happened was of a sudden or catastrophic nature." In other words, it was impossible to tell if Flight 810 suffered some other mechanical failure or weather-related problem before hitting Mount Slesse. There was also no way to tell whether another engine had failed or if fire had swept the wing from No. 2 engine.

The board found that while winds at the upper levels were moving as predicted by forecasters, at lower levels there were erratic, fast-moving crosswinds resulting from Pacific storms that battered the Cascades that night. This, the board proposed, might partially explain how Flight 810 had veered so far off course. With one engine knocked out, the North Star may have begun to yaw (turn on the vertical axis) due to the difference in thrust between the two engines on one wing and the one working engine on the other. This could have caused Flight 810 to veer to the south. "Under normal circumstances," the report stated, "when flying close to the on-course of a radio range such drift would be readily appreciated and adequate correction made to the heading provided the radio signals were good. In this case, however, the aircraft was not close to the on-course signal and such drift would be more difficult to appreciate." The board recommended that in the event of an engine loss, pilots should trim the aircraft for straight and level flight.

There was also a possibility that radio signals were poor that night, but the TCA radio operator at Vancouver told the investigators that "no crash static was audible" on his exchanges with the crew on Flight 810. Three pilots flying that night—Captain Rickard, Captain Wright and First Officer Moir—told the investigators that they did not recall unusually heavy static, but Rickard remembered there being some on lower frequencies. Wright explained that the intermittent loss of such signals would be normal in such bad weather but he had not noted any abnormality that night. But his testimony contradicted one solid piece of evidence. If radio transmissions had been so clear that night, why had Clarke needed to relay a message between air traffic control and Flight 4?

Captain Rickard told the investigators that the radio signal normally found between Vancouver and Hope lacked a reliable guidance signal. The board's report stated, "Under good reception conditions both Vancouver and Princeton ranges are audible at the midway point between them. Under the conditions that existed that evening, it is probable that radio reception of the range signals was zero in the Hope area and positive fixing of position difficult if not impossible . . . This could contribute, and probably did, to Flight 810's crew's unawareness they were so far south of Green 1."

The available evidence underlined that the North Star was at or above 14,000 feet at Hope. The report found, "Flight 810 on its journey out had reported icing and it is probable that sufficient icing would remain on the aircraft to make maintenance of altitude difficult. This would account for the flight's inability to maintain altitude after the loss of No. 2 engine."

Shortly after passing Hope, the North Star had begun a precipitous fall. Between Hope and Silvertip Mountain—just thirty-four kilometres—the plane dropped 6,400 feet. The report painted a picture of Clarke and Boon struggling to control their crippled plane as it dropped through the night sky, too preoccupied even to radio their concerns to controllers in Vancouver. "Precipitation was severe enough to cause a loss of altitude of this magnitude over such a short distance, or . . . [there may have been] some other cause such as the loss of power or encountering similar subsidence to that reported earlier by Flight 4 when on Green 1 west of Hope." The report insisted that while the aircraft was dropping fast, there was nothing to indicate it was free-falling out of control.

Board chairman Desmond Murphy visited the US radar base at Birch Bay and was convinced that according to the last sighting on the screen, Flight 810 had been flying above 10,000 feet, and possibly as high as 14,000, before the blip disappeared. The investigators considered the possibility of yaw, with the North Star being pushed to the south because of two engines working on the right and only one on the left. Pilots are trained to compensate for this, but the investigators speculated that because of possible interference in the electrical system, the instruments may not have reflected how far the plane had swung to the south. From the radio messages it was clear that Captain Clarke was concerned about keeping well clear of the eastbound Flight 4 in order to avoid a mid-air collision, but how did he get so far south?

A concrete answer remained elusive, but the most likely reason was that Clarke and Boon had decided to remain south of the airway until they were sure that Flight 4 was well past them. The report continued, "It is reasoned that the crew of Flight 810 was aware they were south of Green 1, but not to the extent of departure. It is considered probable that more or less extensive interference to the reception of LF radio ranges contributed."

The inquiry board summed up by stating it was impossible to tell precisely why the plane was flying low enough to crash into Mount Slesse but it was probable that the aircraft encountered either severe icing, turbulence, subsidence or a combination of all three.

Even after many months of close examination by leading experts and many witness testimonies, why sixty-two people died on Flight 810 will remain a mystery.

The findings were released on April 28, 1958 and presented to George Hees, the federal Minister of Transport. Out of it came some recommendations for changes in the aircraft signalling system. The investigation revealed that the sighting of the Hope Beacon was unsatisfactory, and the signal should be strengthened or the beacon relocated. This was a side issue, according to the inquiry board; it maintained that the beacon's inadequacy was not in any way a factor in the crash of the TCA plane. The board also urged that consideration be given to setting up procedures that would ensure assistance for all planes from military radar stations, particularly in mountainous areas. The board recommended that the RCAF be approached to assist civil aviation in emergency situations.

20

What If?

The results of the inquiry were frustrating and unsatisfactory for everyone. No one could explain why the return trip to Vancouver ended in that terrible crash. There was no clear reason and no one to blame. Those left behind to grieve could only dwell on the "what-ifs" and all the questions that remained unanswered.

The Department of Transport admitted that it would probably never know what led to Flight 810's fatal crash. For a time everyone tried to move on and forget the terrible tragedy of December 9, 1956, but speculation never really ceased.

Initially there was no criticism of the crew's handling of the plane. In 1970, however, retired TCA president Gordon McGregor publicly disagreed with comments vice-president Herb Seagrim had made at the time of the disaster. Seagrim had strongly supported Captain Clarke's decision to turn back to Vancouver after losing the

No. 2 engine. He had stated that the pilot had followed all the rules and that he would have done the same himself. Nearly fourteen years later, McGregor questioned Seagrim's words in his book *The Adolescence of an Airline*, published by Air Canada in 1970. Noting that Flight 810 had run into a "heavy electrical disturbance" and very high winds, he contended, "One can never know specifically what went into the making of a decision of this kind but it would appear that had the flight continued on an easterly course, with the assistance of the strong west wind, it would have made Calgary quite easily on the remaining three engines." He stressed that the North Star had to fight strong westerly winds after it swung back to the coast, whereas the wind would have assisted the plane if it had continued to fly eastward.

McGregor, a veteran RCAF fighter pilot who had flown in the Battle of Britain, knew all the available details of the fatal flight and probably understood the challenges more than most. It should be noted, however, that general opinion among the aviation world supports Clarke's actions and the practice of heading for the nearest airport in times of trouble.

If the storm had preceded the plane to Calgary and made landing there impossible, Flight 810 would have had to limp onward to Saskatchewan or Manitoba on three engines. Both McGregor and Seagrim knew that Abbotsford Airport in the Fraser Valley was a closer alternative landing strip. It sat about sixty-five kilometres east of Vancouver and was frequently used in emergencies and when Vancouver Airport was socked in with fog.

In the absence of clues among the wreckage of Flight 810, many members of the public still considered the investigation wide open and began asking the big questions themselves. Was Flight 810 doomed from the moment Captain Clarke swung his plane around? Why was the North Star so far south of the Green 1 air lane? Why did the plane descend so rapidly? The biggest question of all: What would have happened if Clarke had been able to maintain an altitude just 100 feet higher as he approached Mount Slesse? It was Elfrida Pigou's first thought after she saw a piece of the plane hanging precariously on a crag at the top of the peak. If Flight 810 could have skimmed over the saddleback top of Slesse, it might have made the return flight safely as the mountains fall away quickly. Climbers who reached the top made varying guesses as to the exact distance between the crash site and the summit.

Slesse is the tallest remaining barrier before the flatlands of the Fraser Valley. Flight 810 had missed Mount Silvertip's towering 8,000-foot (2,500 m) spire thirty-four kilometres to the east. Some guessed that even if Flight 810 had cleared Slesse, it would have struck another of the region's granite peaks. Like many details related to the North Star's crash, all of this was conjecture.

While there was no evidence of major interference in radio communications, there had been trouble clearly understanding at least two messages. It was acknowledged that on a very wild stormy night almost anything was possible and messages could have been lost. Comparing the radio and navigation instruments used a half-century ago and those of today is like comparing a biplane with a space ship. Today, modern computerization and navigational assistance from satellites can take an aircraft over oceans to within a few feet of the centre of a final destination runway. The automatic pilot system takes over much of the flying apart from landing and take-off. Radio interference is extremely rare. The system in the 1950s got most people where they wanted to go safely, but it was simplistic, and atmospheric storms could wreak havoc. The report stated, "Under the conditions that existed that evening it is probable that radio reception in the range signals (those between Princeton and Vancouver) was zero in the Hope area and positive fixing of position difficult, if not impossible."

It will never be known exactly what happened on the flight

deck as the plane turned back. It will never be known whether the pilot was able to tell the passengers what was happening as he tried to control the aircraft.

All of these factors touch the limits of possibility. Fifty years on, any explanation for the crash is circumspect. History has not brought wisdom in this case. What we know now is what we knew then: a dead engine, a stormy night, a wrong course, a loss of altitude, a calm captain, a disaster. The Flight 810 accident remains the worst air crash in western Canadian history and one of the ten worst in the history of the country.

21

TCA Builds a Monument

On Tuesday, December 7, 1957, two days short of the first anniversary of the Mount Slesse disaster, a quiet, reverent crowd stood just off a logging road hemmed in by snow-capped mountains. People bowed their heads, thinking of loved ones who had died so close to this place. It was cold but sunny, vastly different from the wind-packed violence that had raked the region a year earlier when the North Star plunged to its doom. The 350 people who stood so quietly near the base of the mountain could see only Slesse's shadow, not the peak itself, but its presence was felt strongly as wispy clouds moved along the soaring ridges and rays of wintry sunshine broke through to touch the solemn faces of those gathered for the occasion. Poignant words broke the silence. "They passed beyond our touch, beyond our sight, never beyond our love and prayers."

The event was the ceremonial unveiling of a monument that TCA had constructed to honour its three employees and the fifty-nine passengers who died with them on Flight 810. The company flew in relatives and friends of those who had perished. They came from across Canada, from the United States, from Hong Kong and Japan. They had made their way up the Fraser Valley from Vancouver and then travelled by bus from the Chilliwack army camp along the Chilliwack River to the chosen location. The site was not the mountaintop where so many had died, but a quiet clearing on the north side of the Chilliwack River in the midst of the thick coastal forest at the base of Mount Slesse.

The names of the sixty-two victims were engraved on a bronze plaque and attached to an eight-foot-tall granite slab, forming a simple but impressive monument. Mourners ranged from those who had known the two Rowan children—the youngest victims—to the family of Yuen Wah Yoon, the Chinese teenager who had just arrived to start a new life in an unknown land. They included the families of the trio of Calgary women who had been returning home from a wonderful Hawaiian vacation and a subdued group of young women who had taken stewardess training with Dorothy Elizabeth Bjornson. The mourners took turns standing in quiet contemplation before the cairn scanning the list for the names of those they loved.

It was cold and the ceremony was brief. Services were conducted by Father Gordon McKinnon of Chilliwack, Reverend Major Ray Dunford of the Canadian Armed Forces and Rabbi B.A. Woythaler. The band of the Royal Canadian Engineers from Chilliwack played music for the ritual. One grief-stricken mourner became overwrought and was helped away, but the others were quiet and restrained. It was the first time most of them had visited this awesome mountain, a place they would never forget. Somewhere, far above them, Mount Slesse held the remains of their loved ones. Many mourners placed

The eight-foot-tall granite slab with bronze plaque that TCA constructed on Slesse Road to honour its three employees and the fifty-nine passengers who died on Flight 810.

flowers at the base of the granite slab and then silently climbed aboard buses for the journey back.

When the service was finished and a convoy of vehicles began winding its way through the wilderness valley, the passengers glimpsed the sharply etched fangs of the now infamous mountain for the first time. The sun shone through a break in the clouds, and against a clear blue sky glistened the triple peaks, the impact point that took so many lives clear for all to see. At the request of one of the mourners, the procession halted briefly before moving on.

22

A Living Memorial

In addition to the cairn and plaque erected in the shadow of Mount Slesse, there is a living memorial to those who died. Every year since 1957, North Vancouver youngsters have battled in the Gordon Sturtridge Football League. More than 20,000 have participated over the years.

Why is a British Columbia football league named after a Manitoba footballer who played for a Saskatchewan team? Sturtridge played his last game at Vancouver's Empire Stadium in the Shrine All-Star Game. That was the day before he, his wife, Mildred, and four other Western League players died in the North Star disaster. When a new youth football league was being formed on the North Shore—covering the Howe Sound area as far north as Squamish and now Whistler—several of those involved were former prairie boys who had played junior football with Sturtridge. One of them was Stewart McNeill, a former *Vancouver Sun* newspaperman. They knew Sturtridge as a professional player who had volunteered with young footballers as much as he could. They felt naming the new league after him would be a fitting tribute to the all-star, to his fellow players and to the others who died in the North Star disaster.

Today the league is a magnet for players aged six to fifteen and has sent many on to successful careers in the Canadian Football League and the National Football League in the United States.

It is all a vague memory for the second of Sturtridge's three children, Valerie Borthistle, who was only five when her parents died. She and her siblings were raised by their maternal grandparents. Borthistle's Alberta-resident brother Gordon has taken a continuing interest in the Gordon Sturtridge Football League and has attended many events. Valerie Borthistle is proud of the league and feels it is a youthful living tribute to her father, who loved the game of football, and to the memory of all those on Flight 810 that fateful night.

The Canadian Football League instituted an ongoing tribute of its own. Every year the DeMarco–Becket Memorial Trophy—named for Mario DeMarco and Mel Becket, two of the Roughriders players who died in the North Star crash—is awarded to the best lineman in the Western division. The Regina club also retired the jersey numbers of the four players killed: Sturtridge, 73; Becket, 40; DeMarco, 55; and Syrnyk, 56.

23

Preserving the Site

As time passed, summer climbers began frequenting Mount Slesse—some just to see it from afar and others to climb the mountain and view the crash site. Reports of hikers making the arduous ascent and then plundering the wreckage became increasingly common. The stories proved particularly upsetting for relatives, who complained about the situation to local politicians. The annual movement of the ice field, where parts of the plane lay, had continued gradually, sweeping some of the remnants to a lower level where they became more easily accessible to those bent on salvage. A few would-be treasure-hunters still believed they might find the money belt supposedly worn by Kwan Song.

The provincial government had originally intended to protect the area from public intrusion, but it failed to enact legislation to back the plan. The proposal had slipped through bureaucratic cracks.

Maple Ridge resident Andy Cleven, whose uncle Harold Cleven was lost on Flight 810, became incensed by what he saw as desecration. Politicians who had promised the site's protection refused to act. As a result Cleven contacted more than thirty families of other victims and together they petitioned the provincial government, demanding that its promises to designate the mountain as a protected memorial be fulfilled. Their efforts would take time to pay off.

In 1994 the Slesse crash once again became front-page news. A television station reported that some of the crash remains had been stolen from the mountain. A *Vancouver Sun* reporter followed up on the story and interviewed Glen Fetterly, a Chilliwack logger who admitted to taking several items from the site when he made his first climb up Slesse only months after the wreckage was found. He knew the mountain well and had climbed to the lower levels more than once. Just twenty-three at the time, Fetterly had taken a piece of a propeller, a couple of watches, an instrument clock, a set of clip-on pilot's wings from a piece of blue cloth, a roll of bills amounting to thirty-seven dollars and various other small items. The story was picked up and featured in radio, TV and other newspaper reports.

Suddenly Fetterly realized the public did not look kindly on what he had done. He hurriedly refuted any charge of grave robbing, describing what he had as a few odds and ends. He admitted he had been fascinated by the crash scene and had returned several times, but insisted, "They were trinkets. The rubble was four to five feet deep. We could have scrounged all kinds of things, but we didn't. It's funny when you come to a place like that—if you've never been there it's hard to relate to."

Fetterly said he didn't feel right about taking the pilot's wings but explained there were no bodies and therefore no question of stealing from the dead. "I thought at the time, 'It's not mine, but for now I'll take possession.'" He added that if he had wanted to loot the site he would have taken two large bags of mail that he saw there. He often had friends with him when hiking Slesse's slopes and insisted, "I don't want it portrayed that I was prowling around dead people trying to find stuff. We are mountain climbers. We challenge mountains. We hiked a mountain every weekend anyway."

Andy Cleven stepped forward to speak for the victims' families. "We want the area given the same status as any other graveyard," he told reporters. "People can walk through and see the scenery, but we don't want them walking off with what they find." He added that identifiable items should be returned to the families.

Fetterly agreed, and the pilot's wings were turned over to the Clarke family. Public reaction and criticism of the failure of authorities to take action to protect the site continued, but it was still some time before their efforts finally brought results.

24

Major Philip Edwin Gower, MC

The Fetterly controversy erupted right around the time that Max Abrams, a veteran of the Royal Winnipeg Rifles (the "Little Black Devils") became aware of the Slesse monument, which by the mid-1990s was deteriorating badly after years of neglect. After the war Abrams had returned to Winnipeg but later moved west to be closer to his daughter. In Chilliwack, he found several members of his old regiment and they began to meet on a regular basis. The veterans fondly remembered Major Phillip Gower, the man who had led them onto the bullet-swept beaches of Normandy on D-Day and who had been one of the passengers on Flight 810.

Following the June 6, 1944 D-Day attack, the Rifles battled past five concrete casements and fifteen machine-gun positions, according to the Canadian Press's Ross Munro. "They broke into the casements, ferreted out the gun crews with machine guns, grenades, bayonets and knives," Munro wrote. "The Canadians ran into cross-fire. They were shelled and mortared even in the German positions, but they kept slugging away at the enemy. After a struggle that was bitter and savage the Winnipegs broke through into the open country behind the beach."

Max Abrams was Gower's sergeant major that day, one of those lucky enough to remain alive and unscathed. The regimental history states that casualties were heavy but they took and destroyed three casements and twelve machine-gun nests. Major Gower was awarded the Military Cross for his courage and leadership on D-Day. He left the military after the war but rejoined as a member of the Princess Patricia's Canadian Light Infantry and had been on his way home to Winnipeg aboard Flight 810 following a year's duty in the

The Little Black Devils still meet regularly in Chilliwack for coffee. Photographed in 2006, left to right, are Max Abrams, Bob Waite, Bobby Foster, Ed Bogan and Walter Georgeson. Bob Brown photograph.

Max Abrams, photographed in uniform, became the unofficial caretaker of the Slesse monument and still places a wreath each Remembrance Day in honour of his commanding officer, Major Phillip Gower, one of the passengers who died on Slesse.

Far East. He was eager to be home to see his baby daughter for the first time.

Max Abrams decided to look for the monument near the Chilliwack River. It was with some difficulty that he finally found it beside the dirt road, obscured by overgrown trees and bushes. He was appalled by its condition. He offered to do some work on it but was told someone was already looking after it. But as time passed and the monument remained untouched, Max decided to do something about it himself. He purchased a bag of plants from a local nursery and, using two white-painted tires as borders, he fashioned two flower beds at the base of the monument and cleaned it up as best as he could. He then decided he would place a wreath each year on November 11, Remembrance Day, in honour of his commanding officer.

Abrams and a handful of former Little Black Devils still meet in Chilliwack once a week for coffee, their numbers now down to about six. Among them are Bobby Foster, Charlie Gordon and George Hill.

Hill was one of those who heard a loud noise on the night of December 9, 1956, likely the crash in which Gower died.

The monument site would be an utter disgrace but for the efforts of Max Abrams and a few others. It is not much visited now and is barely noticed by passersby. Each November 11, however, until there are no more Little Black Devils to pay their respects, the gallant major who cheated death on the beaches of a French beach only to die on a BC mountaintop will be remembered by his men.

The area is now surrounded by trees that droop depressingly over the monument almost hidden beneath them. The bronze plaque with all the names is dull, dirty and stained with bird droppings. There is little open space left around it.

Air Canada (once TCA) doesn't seem to care. Even when the plaque was erected, the victims' relatives probably didn't notice that the wording followed an unwritten rule of airlines: to detach themselves from disasters as much as possible. The plaque reads, "In memory of the passengers and crew who lost their lives in a North Star aircraft on Mount Slesse Dec. 9, 1956." There is no mention of it being TCA's plane.

25

Archeological Study

I t was a long time before Andy Cleven's group finally forced the government to preserve the crash site. In August 1994, the Ministry of Forests contracted Millennia Research to conduct an archeological study of the area. A final report was to map all crash debris, including any human remains.

Even after so many years, the researchers were moved by the project. Their final report stated, "Conducting the actual fieldwork on Mount Slesse was an emotional experience. Despite the passage of the thirty-seven years since the crash there were often reminders of the human aspect of the tragedy and a frequent flyer with a vivid imagination can feel an empathy with the victims. The mountain itself remains an awesome, powerful place, with its huge, near-vertical walls, frequent rock falls and snow and ice avalanches. It is perhaps the most impressive monument to the human tragedy."

The research team conducted surveys from the air and on the ground. From a helicopter they viewed the impact point near the summit of the third peak as well as other inaccessible areas of the mountain. They confirmed that the distribution of crash debris was limited primarily to the east face of the mountain at an elevation of 5,200 feet (1,600 m). The area had earlier been estimated to be closer to 5,600 feet (1,700 m).

The researchers monitored the area for a seven-day period in September and found that in the years since the crash, powerful avalanches had swept down the east side of the mountain, disturbing the wreckage and carrying debris down to ledges below. The North Star's shattered remains still told the story of the terrible impact, but the debris field was now vastly enlarged. It was altered in many ways, most noticeably because it now included a huge boulder about twenty metres in diameter that had rolled partway down the mountain, perhaps loosened by the force of the crash and then dislodged at a later date by one of the constant summer avalanches. In total, the crash area now covered 582 hectares.

At lower levels researchers found several personal articles and items of clothing, including small shreds of shirts and blouses and a still-knotted tie. Lying in a hole they found part of a black or brown leather flight bag with gold letters that read "Captain A.J. Clarke." It would be the only thing the researchers removed from the site. They later turned it over to the pilot's family, who identified it as a remnant of the leather briefcase presented to Clarke by the cadet corps in Winnipeg.

Grimmer finds included some human remains, mostly parts of arm or leg bones. But after so many years of exposure to the mountain elements, they were difficult to identify. "Much of the unburied human bones are probably in small fragments," a report from researchers stated, "and even the large pieces are very difficult to tell apart from dry wooden sticks." These remains were deposited in a cairn. The researchers also found what was left of the rough-and-ready cairn built in 1957 by Coroner Glen McDonald and RCMP Corporal Tom Anderson.

The archeologists could find no trace of the two burial sites where searchers had entombed the remains of twenty-six people thirty-seven years earlier. The report concluded that the burials had probably been swept away by avalanches, and they quoted one witness who stated that

Sig Peters, one of Chilliwack's enthusiastic hikers, stands beside a sign erected by the Families of Slesse at the start of the trail leading to the crash site.

some body parts interred at McDonald's first service had been swept away about a month later. It had not been possible to bury them deeply enough on Slesse's rocky surface.

The report explained the confusion and complication concerning the intention to safeguard the site as a memorial. The original federal order banning access to the site and approaches by aircraft within three kilometres of the peak had been withdrawn in 1958. Officials were unable to find any trace of an order from the provincial Ministry of Mines and Forests. As a result, in 1994, the Heritage Conservation Statutes Amendment Act was passed providing automatic protection for aircraft wrecks more than two years old.

The new law finally resolved Cleven's complaints. He and other relatives of victims watched from the gallery of the Victoria Legislature on May 29, 1995, as Forests Minister Andrew Petter spoke on the matter.

> I rise today on a matter of significance to British Columbians, to make a ministerial statement, one that concerns the final resting place of sixty-two people who perished in

163

a plane crash on Mount Slesse on December 9, 1956. This tragic event devastated families and friends of the victims and touched sensibilities in every part of Canada. When the scene of the crash was finally discovered, the wreckage was found covering a vast expanse of rock and ice in a high mountain reach. The site was deemed so remote and inaccessible, and the devastation so complete, that no further attempt should be made to bring down the remains and that Mount Slesse should remain their resting place.

The course of time has belied the hope that Mount Slesse would remain a remote and inaccessible wilderness. It has also revealed that an important promise that was made at the time—namely, that this place would be dedicated to the memory of the lives it had claimed—was never actually carried out.

This disturbing fact has renewed the distress of families who believed that a special status had been conferred in recognition of their loss. All members of this House feel compassion for the anguish this has caused the families of Slesse. Today it is our privilege to fulfil a commitment that will give them comfort. On behalf of the Province of British Columbia, I would like to acknowledge the oversight and say to the families of Slesse—many of whom are with us in the gallery today—that we share your wish that this place be dedicated to the memory of your beloved ones.

I am pleased today to announce the creation of a Mount Slesse commemorative site covering approximately 582 hectares on the mountain's east face. The crash site and debris field are now formally recorded as a heritage wreck in the BC archeological site inventory and are thereby protected by statute from damage or desecration. Consistent with the families' wishes that the site be managed for quiet contemplation of its grandeur, a protective zone is now placed around the debris field. Within this area, by means of orders-in-council passed under the Land Act and the Forest Act and by ministerial order under the Mineral Tenure Act, intrusive uses have been prohibited. At a future date modest signage will be placed to identify the nature of this special scene for the hikers and mountaineers who happen upon it. It is our sincere hope that by these actions we can finally lay to rest a troubling memory and thereby initiate a process of reconciliation that is long overdue.

Following the announcement, a plaque was installed on the granite Slesse monument that read:

Mount Slesse Commemorative Site—On December 9, 1956, Trans-Canada Airlines Flight 810 disappeared while on a flight from Vancouver to Calgary. It was not until five months later that climbers found the crash site near the summit of the third peak of Mount Slesse. The destruction of the aircraft was absolute and due to the real dangers of avalanches the area was deemed to be unsafe. Subsequent efforts to remove any bodies or personal effects were abandoned with the belief that a "cemetery in perpetuity" would be created on this remote and inaccessible site. This was not done and the crash site was left unprotected for thirty-eight years.

The plaque also contains a quotation from the book *Cloud Walkers* by newspaperman and mountaineer Paddy Sherman: "This is a place I would gladly lie. The rock is clean and bereft of sadness, the deep silence has a tranquility all its own. At their heads, peaks reach for the sky in a sweep of grandeur, and the mountain itself is their monument."

Vivian Clarke's Resting Place

After her husband's death, Vivian Clarke, the widow of Captain Jack Clarke, lived in the home she had shared with her husband. She devoted her time to raising their two sons and later spending time with her two grandchildren. She never remarried. She died in 2003.

At the time of her death the house still contained many mementos of her husband. Among them was a letter she had received in 1956 from a member of Jack Clarke's Bomber Command crew. The writer recalled a night raid over Germany when their plane was caught in enemy searchlights. While desperately trying to avoid anti-aircraft fire, the heavy Halifax bomber stalled and all four engines quit. As the plane plummeted toward the ground, Clarke managed to get two engines restarted and struggled until a third finally kicked in. They

Vivian Clarke in flight attendant uniform. Used with permission of Jay Clarke.

dumped their bomb load and limped back to England where they landed safe and sound. His performance during the war was one of the reasons Vivian had been so sure her husband would be found safe.

Before her death, Vivian Clarke had expressed a wish to be buried beside her husband, but the government's decision to establish the site as a permanent memorial now made that wish impossible to fulfill. As Jay Clarke—their eldest son and now a Vancouver lawyer and novelist—sat vigil beside his mother's deathbed, he promised to reunite her, at least in spirit, with "the love of her life."

When Vivian died, Jay applied for a special permit to visit Mount Slesse and hired a helicopter. Accompanied by Vivian's brother and a bugler/piper, Jay Clarke carried his mother's ashes to a remote wilderness site with a clear view of the scarred fangs of Mount Slesse. In the quiet of a sunny afternoon, the mountains suddenly rang with a bugle rendition of "The Last Post," and the bagpipes followed with "Flowers of the Forest." After a minute of silence, Jay Clarke laid his mother's ashes to rest beside the area where her husband perished while the strains of "Amazing Grace" echoed back from the mountains like a phantom pipe band from the heavens. It was the end of a great love story.

Epilogue

Another Mountain Claims Elfrida Pigou

Less than three years after she found the remains of Flight 810, Elfrida Pigou prepared to climb one of BC's toughest peaks. She intended to climb "Mystery Mountain," the tag given to Mount Waddington before it had been officially named.

Mount Waddington lies at the head of Knight Inlet, 280 kilometres northwest of Vancouver. For many years it lay hidden behind the solid mass of the coastal range. At 13,104 feet (3,994 m), it is the tallest peak located entirely inside BC. It was first spotted from Forbidden Plateau on Vancouver Island in the 1920s and first climbed in 1927 by Don and Phyllis Munday. It was named for Alfred Waddington, a "champion of noble causes" who arrived in Victoria in 1858 from England. He was involved in BC's early development and attempted to establish a wagon route from the head of Bute Inlet to the booming Cariboo gold fields.

In July 1960, Pigou and her climbing companion Joan Stirling intended to be the first women since Mrs. Munday to scale the magnificent pinnacle. They were accompanied by John Owen, the leader of the party, and Derrick Boddy, another experienced mountaineer. Owen had earlier been a member of the first Canadian party to summit Waddington since the Mundays.

The party of four climbed the shoulder of neighbouring Mount Jeffrey and then proceeded up the Tiedemann Glacier and onto Bravo Glacier, elevation 8,200 feet (2,500 m), where they decided to camp for the night.

The next morning, July 30, they stowed their gear and prepared to pack several loads of equipment up the glacier in relays. Suddenly, an ice-wall located a short distance above the camp collapsed, unleashing an avalanche of ice and snow 140 metres wide and at least 3 metres deep. The four mountaineers and most of their gear vanished. Only one sleeping bag remained.

It was two days before another climbing party appeared on the scene. A large contingent from Seattle had been following the footsteps of the four Canadians. They were surprised and somewhat apprehensive when they saw the tracks disappear into a slide area on the glacier without reappearing on the other side.

The seventeen-person American group was led by Frank Fickeisen, who could not report the accident until his party had completed their climb and were flown out in mid-August. He said, "Several times our people saw the BC party relaying loads up to the flat spot below the ice-cliff." After that they disappeared.

Fickeisen continued, "On Monday, August 1, we followed their tracks up the glacier towards the saddle of Bravo. The snow was very soft and we had no trouble following them. The glacier was broken by crevasses and we passed under an ice-cliff and found an area where the ice had peeled off and covered the flat spot. Down below and out of reach was a sleeping bag. It was midday and warming rapidly." Fickeisen decided it was too dangerous to remain in the area and hurried on as quickly as possible. The next day he and his fellow climbers went higher up Bravo and found a single track that went up and then returned to the avalanche zone before vanishing. Someone from Pigou's party had climbed up there to check conditions and then returned to camp.

The fate of the four Canadians was unknown until the US group returned to Vancouver on August 15. The Mountain Rescue Group, of which Pigou had been such and active member, prepared immediately to fly to the area. Pigou's long-time friends Ian Kay and Fips Broda flew over the site with an RCMP officer and decided there was little hope that any of the four survived the avalanche. By August 20 they decided that there would be no attempt to recover the bodies, which would remain buried on a mountain of ice and snow. It was a fitting resting place for Elfrida Pigou, who had so loved the mountains and glaciers of BC.

At Pigou's death, a close climbing friend of hers would say, "Climbing was her whole existence, and she pursued it with a single-mindedness and strength that few would believe existed in her tiny ninety-five pounds of bone and sinew. She seemed to float up the mountain ahead of the party, kicking steps and breaking trail for friends plodding behind. Her pack held what was needed for any emergency and the little thermos of tea was always forthcoming."

Two years after the accident, the Alpine Club of Canada erected a memorial cairn 5,300 feet (1,615 m) up Waddington bearing the names of the four Canadians.

Today Elfrida Pigou remains very much a part of the spirit of the mountaineering community in BC. On March 28, 1967, a peak in the Ape Lake region of BC's south coast was officially named Mount Elfrida. Pigou had been the first to climb the then-unnamed peak, although she had commented at the time that it was not challenging enough for her. It is ironic that Elfrida Pigou, who found the remains of the worst air crash in the history of western Canada, would die in what Paddy Sherman called "the worst climbing disaster in BC mountaineering history."

Index